Springer Japan KK

Edited by

Japan Association for Evolutionary Economics/Y. Aruka

Evolutionary Controversies in Economics

A New Transdisciplinary Approach

With 65 Figures

Springer

Japan Association for Evolutionary Economics

Yuji Aruka
Professor of Economics
Faculty of Commerce
Chuo University
Higashinakano, Hachioji
Tokyo 192-0393, Japan

ISBN 978-4-431-67994-3

Library of Congress Cataloging-in-Publication Data

Evolutionary controversies in economics: a new transdisciplinary approach / Japan Association for Evolutionary Economics; Y. Aruka (ed.).
p. cm.
Includes bibliographical references and index.
ISBN 978-4-431-67994-3 ISBN 978-4-431-67903-5 (eBook)
DOI 10.1007/978-4-431-67903-5
1. Evolutionary economics. I. Aruka, Yuji, 1949- II. Japan Association for Evolutionary Economics.
HB97.3 .E94 2001
330—dc21 2001020958

Printed on acid-free paper

Originally published by Springer-Verlag Tokyo in 2001
Softcover reprint of the hardcover 1st edition 2001

Typesetting: Best-set Typesetter Ltd., Hong Kong

SPIN: 10780783

Preface

In March 1997, we launched the Japan Association for Evolutionary Economics (JAFEE) to gather the academic minds that, out of dissatisfaction with established dynamic approaches, were separately searching for new approaches to economics. To our surprise and joy, as many as 500 members, including graduate students, joined us. Later that year Prof. Horst Hanusch, then President of the International Joseph A. Schumpeter Society, remarked that such a start would take a couple of decades in Europe to prepare for. Since then we have been developing our activities incessantly not only in terms of the number of members, but also in terms of the intensity of international academic exchange.

Originally the planning of this book came about as the successful outcome of our fourth annual conference organized as an international one, JAFEE 2000. Incorporating other international contributions related to our preceding conferences, this book has eventually turned out to be one of the most enterprising anthologies on evolutionary economics ever published. Specifically, it contains excellent papers on such topics as streams of evolutionary economics, evolutionary nonlinear dynamics, experimental economics and evolution, multiagent systems and complexity, new frontiers for evolutionary economics, and economic heresies. In short, this book will provide a vivid and full-fledged picture of up-to-date evolutionary economics.

I very much appreciate the valuable contributions to the book by the following overseas professors: Richard Nelson (Columbia University, USA), Ulrich Witt (Max Planck Institute, Germany), Geoffrey Hodgson (University of Hertfordshire, UK), Tönu Puu (University of Umeå, Sweden), Daniel Friedman and Nirvikar Singh (University of California, Santa Cruz, USA), Shu-Heng Chen (National Chengchi University, Taiwan), Stephen J. Guastello (Marquette University, USA), Domenico Delli Gatti (Catholic University, Italy), Mauro Gallegati (University of Teramo, Italy) and Tony Lawson (University of Cambridge, UK). I also offer my special thanks to Springer-Verlag Tokyo for their sincere cooperation in publishing this book as well as the three volumes of consecutive publications JAFEE has planned as one of its important projects.

Satoshi Sechiyama
President of the Japan Association for Evolutionary Economics
April 9, 2001

Introduction to JAFEE

In the late twentieth century we perceived a positive movement toward innovating economics from a new wave of evolutionary strands. Deterministic ideas on economic systems prevailing in the established mainstream should be reconsidered, we felt. Assurance of our conviction was greatly due to the effects of enlightenment from the work of Sydney Winter and Richard Nelson. We had seen the founding of several societies such as the European Association for Evolutionary Political Economy and the International Joseph Schumpeter Association by the late 1980s, establishing currents of evolutionary thinking in economics by promoting publication of journals both in Europe and in the United States. These worldwide phenomena were encouragement enough for us to organize a new society called the Japan Association for Evolutionary Economics (JAFEE). At the inaugural meeting in Kyoto, the old capital of Japan, in March 1997, one of the guests was Richard Nelson. You might think of this organization as being just a latecomer. Yes, it is, in terms of chronological order in the evolutionary stream mentioned above. In view of the new features of the organization, however, it is definitely not a laggard.

It is true that on the one hand our organization shares many common features with our antecedents, deeply committed to evolutionary ideas originating with Veblen, Schumpeter, and the Austrian school. On the other hand, our society also has a strong interest in empirical studies on technological and institutional developments in modern society, in particular, in sympathy with Nelson and Winter. These efforts contribute to a development of nonformalistic analysis taking into account the "process reality" of our economic systems. But these approaches do not constitute the whole picture. According to Nelson's taxonomy, there are at least three distinguishable subgroups, evolutionary game theory, nonlinear dynamic systems, and variation–selection models, underlying the ideas of evolutionary economics. These subgroups' activities can never be separated from the frontier sciences arising in the late twentieth century. Those disciplines have led to a new approach to integrate social and natural sciences, together with computer sciences, i.e., a transdisciplinary approach, with the eventual aim of a comprehensive or appreciative theory of economics. It is noteworthy that our organization is equally concerned with such a new approach.

JAFEE thus is constituted of certain diverse subgroups whose particular interests range from evolutionary and institutional economics to complexity sciences. These groups independently are organizing workshops throughout the year. We regularly hold an annual meeting as well as an autumn conference in the style of a symposium. Our meetings from the first have not been limited to domestic concerns. JAFEE 2000 deserves special mention for its international tone.

From the very outset of our activities, we have established an editorial function to issue a newsletter, and we now are bringing out a series of books in Japanese on evolutionary economics titled *Genesis Series of Evolutionary Economics*, to be published by Springer-Verlag Tokyo. Concurrently, we are also interested in publishing English titles such as *Evolutionary Controversies in Economics*.

JAFEE is governed by an executive board made up of some thirty members, and currently is administered by a secretariat located for the time being at Kyoto University, where our founding meeting was held.

Office contacts: Professors Kiichiro Yagi and Kazuo Yoshida, Graduate School of Economics, Kyoto University, Yoshida-Honmachi, Sakyo-ku, Kyoto 606-8501, Japan

Web site: http://www.econ.kyoto-u.ac.jp/evoeco

President: Professor Satoshi Sechiyama, Department of Sociology, Kansai University, who is also Professor Emeritus, Kyoto University

Vice President: Professor Yoshinori Shiozawa, Faculty of Economics, Osaka City University

Yuji Aruka
Editorial Board, The Japan Association
for Evolutionary Economics

Contents

Part VI Economic Heresies

Contributors

Aruka, Yuji (Chairperson, JAFEE2000)
Professor, Faculty of Commerce, Chuo University, Higashinakano, Hachioji, Tokyo 192-0393, Japan
aruka@tamacc.chuo-u.ac.jp

Nelson, Richard R.
Professor, School of International and Public Affairs, Columbia University, 420 West 18th Street, New York, NY 10027, USA

Hodgson, Geoffrey M.
Research Professor, The Business School, University of Hertfordshire/Director, Centre for Research in Institutional Economics, Hertford Campus, Mangrove Road, Hertford, Hertfordshire SG13 8QF, UK
g.m.hodgson@herts.ac.uk

Witt, Ulrich
Professor, Max Planck Institute for Research into Economic Systems, Executive Director, Evolutionary Economics Unit, Kahlaische Str. 10, D-07745 Jena, Germany
witt@mpiew-jena.mpg.de

Puu, Tönu
Professor, Department of Economics, University of Umeå, SE-90187 Umeå, Sweden
Tonu.puu@econ.umu.se

Asada, Toichiro
Professor, Faculty of Economics, Chuo University, Higashinakano, Hachioji, Tokyo 192-0393, Japan
asada@tamacc.chuo-u.ac.jp

Guastello, Stephen J.
Professor, Department of Psychology, Marquette University, PO Box 1881, Milwaukee, WI 53201-1881, USA
6155guastell@marquette.edu

Friedman, Daniel
Professor, Department of Economics, Social Sciences 1, University of California at Santa Cruz, Santa Cruz, CA 95064, USA
dan@cats.ucsc.edu

Singh, Nirvikar
Professor, Department of Economics, Social Sciences 1, University of California at Santa Cruz, Santa Cruz, CA 95064, USA
boxjenk@cats.ucsc.edu

Chen, Shu-Heng
Professor, AI-ECON Research Center, Department of Economics, National Chengchi University, Taipei 11623, Taiwan
chchen@nccu.edu.tw

Delli Gatti, Domenico
Associate Professor, Faculty of Economics, Catholic University, Largo A. Gemelli 1, 20123 Milan, Italy
delliga@mi.unicatt.it

Gallegati, Mauro
Professor, Faculty of Law, University of Teramo, via Memmingen, Colleparco, 64100 Teramo, Italy
gallegati@deanovell.unian.it

Yamashita, Tomohisa
PhD. Student, Laboratory of Harmonious Systems Engineering, Division of Complex Systems Engineering, Institute of System and Information Engineering, Graduate School of Engineering, Hokkaido University, Kita 13 Nishi 8, Kita-ku, Sapporo 060-8628, Japan
tomohisa@complex.eng.hokudai.ac.jp

Ohuchi, Azuma
Professor, Graduate School of Engineering, Group of Information, Electronics and Systems Engineering, Hokkaido University, Kita 13 Nishi 8, Kita-ku, Sapporo 060-8628, Japan
ohuchi@complex.eng.hokudai.ac.jp

Sato, Hiroshi
Research Associate, Department of Computer Science, National Defense Academy, Hashirimizu 1-10-20, Yokosuka, Kanagawa 239-8686, Japan
hsato@nda.ac.jp

Koyama, Yuhsuke
PhD. Student, Graduate School of Economics, Kyoto University, Yoshida-Honmachi, Sakyo-ku, Kyoto 606-8501, Japan
yus@mal.seikyou.ne.jp

Deguchi, Hiroshi
Associate Professor, Graduate School of Economics, Kyoto University, Yoshida-Honmachi, Sakyo-ku, Kyoto 606-8501, Japan
deguchi@econ.kyoto-u.ac.jp

Kurumatani, Koichi
Senior Researcher, Cyber Assist Research Center (CARC), The National Institute of Advanced Industrial Science and Technology (AIST), Aomi 2-41-6, Koto-ku, Tokyo 135-0064, Japan
k.kurumatani@aist.go.jp

Shiozawa, Yoshinori
Professor, Faculty of Economics, Osaka City University, Sugimoto 3-3-138, Sumiyoshi-ku, Osaka 558-0022, Japan
shiozawa@econ.osaka-cu.ac.jp

Takayasu, Hideki
Senior Researcher, Sony Computer Science Laboratories, Higashigotanda 3-14-13, Shinagawa-ku, Tokyo 141-0022, Japan
takayasu@csl.sony.co.jp

Tabata, Minoru
Associate Professor, Department of Applied Mathematics, Faculty of Engineering, Kobe University, Kobe 657-8501, Japan
mnrtabata@wombat.or.jp

Eshima, Nobuoki
Professor, Departement of Medical Information Analysis, Faculty of Medicine, Oita Medical University, Oita 879-5593, Japan
eshima@oita-med.ac.jp

Takagi, Ichiro
Professor, Department of Management Engineering, Faculty of Engineering, Kyushu Tokai University, Kumamoto 862-8652, Japan
takagi@bm-l.ktokai-u.ac.jp

Yoshida, Masaaki
Associate Professor, Faculty of Economics, Senshu University, Higashi-Mita 2-1-1, Tama-ku, Kawasaki 214-8580, Japan
yoshida@isc.senshu-u.ac.jp

Lawson, Tony
Lecturer, Faculty of Economics and Politics, University of Cambridge, Sidgwick Avenue, Cambridge CB3 9DD, UK
Tony.Lawson@econ.cam.ac.uk

Introduction: Toward a New Transdisciplinary Approach for Evolutionary Controversies

Yuji Aruka

Key words. Philosophy, Nonlinear dynamics, Multiagent system, Experimental economics, Inter- and trans-disciplines

1 JAFEE2000 and the Advent of this Book

The Japan Association for Evolutionary Economics, JAFEE,[1] was started at Kyoto University in March 1997, and adopted the slogan "From determinism to pluralism." Evolutionary processes in the domain of social and/or economic beings in essence favor *pluralism* as the best way to elucidate various aspects of the past and future developments of their interests. Thus the association declared: "Accordingly any branch of academism which is consistent with Pluralism is to be equipped with such intellectual framework that incorporates the concept of endogenous evolution and stochastic mutation. Evolutionary Economics is an effort to innovate new economics by delving into the evolutionary mechanism of systems, organizations and technology."[2]

1.1 JAFEE2000

At the founding meeting of this organization, Richard Nelson addressed an inaugural lecture, and his remarks have guided the continued expansion of the association's broad international activities. After three successful annual meetings, it was decided that the fourth annual meeting, on March 25 and 26, 2000, would be the first international meeting, and would be held in Tokyo to celebrate the start of the new millennium. This international conference, called JAFEE2000, was to be incorporated as an extended session in addition to the usual annual meeting for domestic participants. The meeting consisted of seven lecture sessions, 2-day poster sessions, and a panel discussion devoted to the main theme of the meeting, i.e., inter- and transdisciplines for evolutionary economics.

Faculty of Commerce, Chuo University, Higashinakano, Hachioji-shi, Tokyo 192-0393, Japan

[1] Note that JAFEE, the abbreviation of this organization, is the same as that of a different organization, The Japanese Association of Financial Econometrics and Engineering.

[2] Invitation to JAFEE, March 1996. See the official site of the Japan Association for Evolutionary Economics at http://www.econ.kyoto-u.ac.jp/~evoeco/index.html

1.2 Towards Transdisciplines

In preparing this meeting, it was agreed that "its major goal was to bring together researchers from various fields and various places and to encourage transdiscipline among the many studies involved in economics and the related sciences."[3] Here, "transdiscipline"[4] is a new term officially adopted by the Graduate School of Frontier Sciences, University of Tokyo, in order to indicate such fusions of different disciplines. Beneath the ideas of evolutionary economics, there are at least three distinguishable subgroups of evolutionary game theory, nonlinear dynamic systems, and variation–selection models, as Nelson has suggested.[5] The subject areas for this conference were chosen in order to combine these three themes. Thus, the sessions for the international part of the conference were entitled Evolutionary Nonlinear Dynamics, Complexity in a Multiagent System, and Experimental Economics and Coevolution. The web site attracted international attention to JAFEE2000, and many delegates attended the conference from overseas.[6]

1.3 The Advent of this Book

JAFEE2000 clearly was a good opportunity for an international publication about JAFEE. The main part of this book consists of the entire collection of keynote lectures not only by guest speakers for JAFEE2000, e.g., Tönu Puu (Umeå University, Sweden), Daniel Friedman (University of California, Santa Cruz, USA), Shu-Heng Chen (National Chengchi University, Taiwan), and Tony Lawson (University of Cambridge, UK), but also those by overseas visitors at the first and second annual meetings, e.g., Richard Nelson (Columbia University, USA), Ulrich Witt (Max-Planck Institute, Germany), and Geoffrey Hodgson (University of Hertfordshire, UK, and formerly of the University of Cambridge). Needless to say, the speakers at the earlier meetings contributed toward establishing a general overview of evolutionary economics which was philosophically different from established institutional views, and included some applications of biological Darwinism to economics. It is hoped that the contributions

[3] See the call for papers for JAFEE2000: http://arukalab.tamacc.chuo-u.ac.jp/~jafee2000/

[4] In our autumn conference on September 18, 1999, prior to the annual meeting of JAFEE2000, Mitsuhiro Toriumi, one of the main founding members of the Graduate School of Frontier Sciences, University of Tokyo, was invited to describe the founding of the institution. "[W]e have to break the current old discipline and then to reorganize it in order to create a completely new framework of research programs. This might be said audacious challenge against the trend in current academies where research activities tend to be engaged in a quite narrow spectrum of a limited science. However, the challenge is exactly what we will dedicate in the School of Frontier Sciences. The keyword of this new school is 'transdiscipline' standing over research and education programs of current separate science." See the home page of the Graduate School of Frontier Sciences, University of Tokyo, which started to accept regular students from April 1999. http://www.k.utokyo.ac.jp/e/overview.html#overview

[5] See R. Nelson, "A Viewpoint of Evolutionary Economic Theory," in this volume.

[6] The Third Experimental Economics Conference (Chairperson Yuji Aruka) on March 24, 2000, was held immediately before JAFEE2000, Shaym Sunder, from Yale University, gave the plenary lecture. Another meeting was arranged for the second CRUGE (Chuo University) International Symposium 2000, which was attended by Michael Sonis, Bar-Iran University, Israel, Willi Semmler, Bielfeld University, Germany, and others from overseas.

of R. Nelson, G. Hodgson, and U. Witt will be seen as a basic introduction to evolutionary ideas in social sciences. However, this book is not simply the proceedings of JAFEE2000, although it greatly depends on the materials and communications provided by JAFEE2000 activities, since those activities were closely related to the main theme of transdisciplines. It was also fortunate that JAFEE2000 created an opportunity to introduce Stephen Guastello (Marquette University, USA), a psychologist, who is the chief editor of a journal titled *Nonlinear Dynamics, Psychology and Life Sciences* (NDPLS), which will publish a special issue on JAFEE2000. Thus, this book is an integrated report on both the philosophical[7] and the new scientific, analytical aspects of current economic thinking.

2 Preliminaries for Evolutionary Controversies

2.1 A New Transdisciplinary Study

Biophysics has been an established discipline for a long time. We have recently seen the emergence of the new discipline of econophysics,[8] which may be regarded as a typical example of transdisciplinary thought. In this case, we have a new discipline *transformed* from physics, in particular statistical physics, into economics. However, it is rare to find econophysics being regarded as an independent discipline in the way that has happened with biophysics. A transdisciplinary study is usually organized in a loose or temporary association of related inter- and cross-disciplinary fields to solve various challenging contemporary problems. A main theme of this book is to show how transdisciplines can work. We look heuristically at a selection of the subjects chosen in evolutionary economics. It should also be noted that one heuristic selection will consider one temporary style of transdisciplinary study, while another will rectify this style.

2.2 A Heuristic Sample for Transdisciplines

In JAFEE2000, the theoretical inter- and transdisciplines considered are within the limited fields of nonlinear economic dynamics, experimental economics, and multiagent systems. In order to examine transdisciplinary work in those fields, the editor has taken an example from his personal experience. His chapter on the Avatamsaka game[9] considers a social dilemma in coordination which all drivers experience when they decide to give way to others on the road. This is a good example of a transdisciplinary study, since this game can be treated as an evolutionary game in terms of a system of nonlinear differential equations, or as a study in experimental economics using human subjects, or as an example of the Polya urn process with a stochastic model in a multiagent system, which also takes certain interdisciplinary considera-

[7] The philosophical aspects are represented by an interesting application of nonlinear dynamic techniques by T. Puu, and scientific or economic heresies such as the critical realism of T. Lawson.

[8] The first textbook of this discipline was written by Mantegna and Stanley (2000).

[9] Avatamsaka game structure and experiment on the web (p. 113–130). See also Aruka, 2000.

tions about cultural and psychological factors into account (Polya and Eggenberger 1923; Aruka 2001). Several other examples of transdisciplinary studies will be found throughout the following chapters.

2.3 Evolutionary Controversies

During the late 1960s and the early 1970s, there was a big debate about capital controversies between the Massachusetts Cambridge in the USA and the post-Keynesian Cambridge in the UK (Harcourt 1972; Harcourt and Laing 1971). Such a controversy often means a fight for legitimacy between the two conflicting doctrines. In these debates about capital, the legitimacy of the neoclassical doctrine was considered to be vitally important. Nowadays, however, debates about alternative legitimacies are very rare. We are used to considering different competing techniques in the same arena, and other controversies are important. During the late twentieth century, economics encountered several rival disciplines, and the multiagent system is a typical example.[10] In this volume, we adopt competing techniques such as the nonlinear chaotic evolutionary system, the experimental method, or the genetic algorithm (GA) to illustrate a problem in marketing, for example, and consider whether they are complementary or antagonistic to traditional theories. Chaos theory and fractals, as well as the new sources of nonlinearity, for instance, have been sufficiently generic for many scientific discipline to develop their own applications. These techniques and ideas are all relevant to evolutionary processes. Thus, we shall examine evolutionary meanings by evaluating how well competing models perform in the same arena, and/or how different techniques can be generic for applications in various disciplines.

2.4 Market, Utility, and Decisions

The central issue for the debates on capital was capital in aggregate, whose attribute was a kaleidoscopic one, i.e., it was simultaneously both the physical capital and the value of that capital. In evolutionary controversies, kaleidoscopic terminologies of the most elementary usages must be noted. In textbooks, the market still relates to consumer goods, while the market frequently cited in newspapers and magazines referrs to the financial/or brokers stock market. The concept of utility can also be kaleidoscopic. The way of thinking about ordinary utility in terms of preference ordering of a basket of physical goods has been adapted to mean the expected utility of the probability distributions of a set of economic events[11] as an apparently different domain from a traditional physical set of goods. The utility functions of cultural and psychological elements could not usually be considered in the main arguments. However,

[10] The multiagent system approach, by means of its code of artificial intelligence, often argues the possibility of replacing all the main results, and the key concepts such as rationality, of traditional economics with its own alternatives (see, for example, Weiss 1998).

[11] Expected utility has now become much more familiar than the ordinary classical utility function. However, the underlying axioms for expected utility have been seriously controversial ever since its appearance. The most popular set of assumptions was established by J. Marshack in Oxford in 1950, and generalized by P.C. Fishburn in 1982. However, we are faced with several grave problems about the construction of expected utility, since M. Allais, in 1953, suggested an objection to it. These controversies were one of the forces driving the development of experimental economics.

many decisions made in the history of human societies cannot be isolated from some cultural and/or psychological features of the relevant agents, as demonstrated by Robert Frank.[12] The implication of rationality may differ from doctrine to doctrine. Moreover, it is interesting to see that we can have a project to replace the familiar concept of utility with the psychological value for decision making analytically defined by Helbing (1995), although he called it utility (see footnote 26).

2.5 Toward an Empirically Testable Analysis

Studies on auctions are frequent in experimental economics (Kagel and Roth 1995), but in many cases an auction is discussed as a very special market in which efficiency is justified only within a limited range and by way of Marshallian partial equilibrium analysis of economic surplus.[13] Auctions are traders markets irrespective of consumer utility in general equilibrium. Although the law of the brokers market is obviously irrelevant to the law of demand in a consumer market, at least in traditional doctrines,[14] a certain deep faith on the part of the market itself provided us with a general belief in a market law common to both markets in spite of their qualitatively different bases. At present, however, we are far from such a generalization, because the analytical tools currently employed are too poor to achieve this kind of ambition. Werner Hildenbrand (1994)[15] discovered that our general utility function on real goods does not work well enough to guarantee the existence of the law of demand without imposing the Mitjushcin and Polterovich condition.[16] If we need to refer to a law, we should design a model which would be empirically testable by employing several profound null hypotheses, as Hildenbrand has shown. We should never slavishly employ any market law, either in general or particular, unless it is empirically testable.

2.6 Economics of Complexity

As well as the monumental work on evolutionary economics by Richard Nelson and Sydney Winter (1982), there are a few other important publications on the complex economic systems in Japan. One is on the micro–macro loop method of studying

---

[12] See Frank 1988. Frank is famous as a co-author of *The Winner-Take-All Society*, with P. Cook, in 1995.

[13] The fundamental objection to the method of partial equilibrium analysis was put forward by Piero Sraffa in 1925.

[14] When we refer to efficiency, we usually employ the utility function in some form in which the consumer market is supposed to be proven to perform best in serving the community. The myopic pursuit of the self-interest of each agent may guarantee social efficiency in this setting. Brokers in the trade markets never behave according to a traditional utility function, and use a function with a base which has no classical content. Note, however, that Hicksian laws on relative price movements hold without an individualistic utility function, but such price rules mean efficiency necessarily do not. See Morishima, 1983, Chap. 2, for Hicksian laws in a trader's market.

[15] In this book, Hildenbrand suggested a *macroscopic microeconomics*, which is based on the heterogeneity of household income distribution, instead of the traditional microscopic microeconomics, which is based on a presumption of the homogeneity of individualistic utility functions.

[16] A utility function $u(z)$ must be restricted by the condition

$$-z\left[\sum_{j=1}^{l}\frac{\partial^2 u^i(z)}{\partial z_j^2}z_j\right]\Big/ z\left[\frac{\partial^2 u^i(z)}{\partial z_j}\right]<4$$

evolutionary economics proposed by Yoshinori Shiozawa (Osaka City University, Japan).[17] Another is on the holonic loop method, by Hiroshi Shimizu (formerly of the University of Tokyo, Japan),[18] a biophysicist, who, in the early 1980s, developed an experimental computer called the holonic computer to use the semantic loop system. These ideas are promising tools for analyzing a newly specified set of key terms {path dependency, lock-in, increasing returns}. However, these terms are simply an extension of the international reputation of the Santa Fe Institute in the field of the science of complexity in economics. In September 1987, the Santa Fe Institute made the first bold attempt to unite economists and physicists in one discipline at a conference where B. Arthur was impressed by Holland's genetic programming, as S.-H. Chen stressed (Anderson et al. 1988). Recently, the development of this trend has led to the emergence of a new empirical science called econophysics, promoted by H. Eugene Stanley (Boston University, USA) and Hideki Takayasu (Sony Computer Science Laboratories, Japan).[19] This now embraces more tangible empirical analyses for practical purposes. These studies will develop new descriptions of social interaction in different aspects of observation.

2.7 Future Developments

We can expect the development of more new frontiers in evolutionary economics. It is interesting to note that there may exist a branch point dividing chaotic dynamic systems from gradient systems,[20] and that the latter could generate a martingale. Brian Arthur, a Santa Fe economist, referring to a well-known stochastic model of path dependency, found this when he employed a generalized theorem for Polya's urn process.[21] A particular set in this area will certainly be expanded not only toward a chaotic dynamic model, but also usefully toward a stochastic model. As another exemplification of the reinforcement mechanism, a more extensive use for the master equation may be found in an evolutionary spatial economic system. In addition, economic experiment will be expanded to contain a new eclectic of artificial simulations and human experiments, as in the U-mart experiments in Japan, in which Association members such as Y. Shiozawa and Akira Namatame (National Defense

in order to confirm an ever-positive income effect in the Slutsky equation, which could ensure the law of demand at the cost of parting with a general utility function. (See Hildenbrand 1994, p. 18 and Appendix 4; The monotonicity of individual demand functions.)

[17] Shiozawa is the Vice-President of the Japan Association for Evolutionary Economics (JAFEE). He has written many books on economic complexity in Japanese. For an English version, see Shiozawa 1999.

[18] H. Shimizu is Professor Emeritus of the University of Tokyo. The first holonic theory was proposed by Arthur Koestler. The holonic concept suggested by Shimizu is quite different, and is also completely independent of the ideas of Koestler.

[19] They organized an international conference titled: The Symposium on Empirical Science of Financial Fluctuations, November 15 to 17, 2000, held at NIKKEI, Tokyo, Japan.

[20] See, for example, Arthur 1994, p. 41: "Now, a deterministic system like (5) [in Arthur's book] might not itself converge—it might give rise to limit cycles or other more complicated asymptotic behavior. However, we can simplify matters by restricting (5) to a *gradient system*."

[21] The general theorem for Polya's process was originally proven by Yuri Ermoliev and Yuri Kaniovski, in the Ukraine. It is interesting to note that this theorem is the decisive one which proved path dependency, which is now a Santa Fe patent (see Arthur 1994, Chaps. 3, 7, and 10).

Academy, Japan) are deeply involved, and also the V-mart in the USA, and these experiments will lead to a new computer science for economics.

2.8 Philosophical Methods Inspection

Institutionalist thought is one of the three main sources of evolutionary economics theories.[22] There are some economists who do not accept these theories, and it is interesting to see that even an institutionalist such as G. Hodgson should have received severe radical criticism. An approach such as that of critical realism, led by T. Lawson,[23] a Cambridge economic philosopher, may be required to make a balanced judgment of the performance of the doctrines in current use, which we call heterodox economics.[24] Evolutionary economics should be criticized in its own territory. Embracing heretical ideas seems to us to be indispensable to keeping the scope of economic activities evolving.

3 Chapter Profiles

3.1 Perspectives of the Book

All the chapters in this volume have an abstract and five key words (except for that by R. Nelson), from which the reader can assess the main points of each one. The book is divided into six parts, as follows: I, Streams of Evolutionary Economics; II, Evolutionary Nonlinear Dynamics; III, Experimental Economics and Evolution; IV, Multiagent Systems and Complexity; V, New Frontiers for Evolutionary Economics; VI, Economic Heresies.

3.2 Streams of Evolutionary Economics

This section contains the philosophically elementary views of the three authors, who all have a reputation in different streams of evolutionary economics. There was a clear historical movement toward evolutionary economics in the late twentieth century. The chapter by G. Hodgson (University of Hertfordshire, UK) uses statistics to demonstrate the extent to which the pioneering book by R. Nelson and S. Winter has influenced economics as a whole during the last 20 years. Nelson (Columbia University, USA), who is seriously critical of the formalist analytical method, thinks "it important that the process built into that model be at least metaphorically consistent with what is known about the actual processes being modeled." He thus proposes that evolutionary economics must have an "appreciative theory" to adopt "process realism" in economics. The similar idea in the context of institutionalist arguments came from G. Hodgson, who emphasized the need for liberation from formalist strains. Thus, the new approach toward an empirically testable model should be construed on the

[22] See the articles by R. Nelson and G. Hodgson in this volume.

[23] This approach is supported by the group called the Cambridge Realist Workshops, which held a conference entitled "Critical Realism in Economics: What Difference Does It Make?" Conference 2000, May 5–7, Sidgwick Site, Cambridge, sponsored by the *Cambridge Journal of Economics*.

[24] A Heterodox Economics Conference was held on June 28 and 29, 2000, in Manchester.

basis of these reflections on formalistic analysis. According to U. Witt (Max Planck Institute, Germany), we could argue that human economic behavior is predestined in the same terms as Darwinian biology, and that this supersedes the preference subjectivism of formalistic analysis.

3.3 Evolutionary Nonlinear Dynamics

Darwinian thought can still suggest new economic insights. The evolution of a tree may be applied to the bifurcations of economic complexity. Using this idea, T. Puu (Umeå University, Sweden), a JAFEE2000 keynote speaker on nonlinear economic dynamics,[25] gives us an appreciative theory of nonlinear dynamics in relation to the characteristics of change accompanying cultural development without using any mathematical equations. This provides a new philosophical insight into the study of the nonlinear strand of evolutionary economics. There exist several schemata for nonlinear dynamic models which are clearly different from those in the neoclassical stream. An alternative and generally reliable source is post-Keynesian dynamics. Based on M. Kalecki's principle of increasing risk, T. Asada (Chuo University, Japan) formulated a nonlinear financial dynamics of debt and capital to illustrate the Minsky cycle. We also introduce another interesting application of nonlinear techniques. S. Gaustello (Marquette University, USA), an active applied psychologist, who is interested in nonlinear systems as a whole, analyzed the bifurcation structure in the dynamic Philips curve on the relationship between unemployment and inflation rates. This chapter also uses nonlinear regression coupled with structural modeling.

3.4 Experimental Economics and Evolution

One of the most popular subjects in experimental economics is the coordination game, essentially a two-person game but with *n* agents participating, as specified, to face a social dilemma. The psychological elements in the expected utility cannot usually be taken into account, as mentioned in the previous section, but the ideas of Robert Frank are helpful when considering such elements. D. Friedman (University of California, Santa Cruz, USA), a JAFEE2000 keynote speaker on experimental economics, together with N. Singh (University of California, Santa Cruz, USA) have made an interesting attempt to introduce a vengeance factor into their experimental coordination game. Using the Dawkins meme, and referring to the competition among memes that "prescribe human behavior towards culprits outside the group and towards group members who deviate from that prescription," they then argued for their evolution. However, the coordination game is rather restrictive in the sense that two special conditions of complementarity and positive spillover must be satisfied. Y. Aruka (Chuo University, Japan), on the other hand, also examined some similar psychological effects, such as the feeling of revenge, in his original critical Avatamsaka game, which has an infinite number of equilibria because the property of complementarity is not fulfilled. This game is not only formulated in terms of an evolutionary game, but is also achieved in an experiment with human subjects. Interestingly, this game can also be conducted on the web by way of Aruka's original program.

---

[25] In JAFEE2000, in essentially the same spirit, he presented part of his new book (see Puu 2000).

3.5 Multiagent Systems and Complexity

The coordination games mentioned above normally are interpreted as multiagent systems. The genetic algorithm (GA) of J. Holland (1975) not only influenced R. Lucus, a neoclassical economist, but also B. Arthur, a Santa Fe economist. Before presenting some analyses of types of multiagent system, we have a remarkable general review of the relevance of genetic programming (GP) for artificial adaptive agents. This is summarized by S.-H. Chen (National Chengchi University, Taiwan), a JAFEE2000 keynote speaker on multiagent systems, whose interest is in an evolving population of decision rules, i.e., the automatic generation or deletion of rules. One shortcoming of the classifier system is that it lacks this mechanism. Chen also refers to his method of employing GP to discover the underlying laws of motion for some simple chaotic time series. We then have a successful application of the GA to economics, which is provided by D. Delli-Gatti and M. Gallegati (Catholic University, Milan, and the University of Teramo, Italy, respectively), two Italian economists working in nonlinear dynamics. They used a simple model of N firms and L workers with a post-Keynesian flavor to study "the dynamics of the coordination efforts of a fixed number of heterogeneous agent whose choice are modeled through genetic algorithm [of cell automata] in a setting of bounded rationality and limited information." Another analysis of a type of multiagent system presupposes a new type of coordination game called the distributed social dilemma, in which players have discretion in the choice of which game they will play. Suppose, in addition, the presence of metaplayers who can punish other players for their errors. In this experimental setting, T. Yamashita and A. Ohuchi (Hokkaido University, Japan) simulated the effects of competitive metaplayers.

3.6 A New Frontier for Evolutionary Economics

The idea of the U-mart auction design in a virtual stock market on the network is, so to speak, a transdisciplinary project covering computer science, artificial intelligence, and experimental economics in the sense that the agents are both human traders and computer programs. H. Sato, Y. Koyama, K. Kurumatani, Y. Shiozawa, and H. Deguchi (National Defense Academy, Kyoto University, Electrotechnical Laboratory, Osaka City University, and Kyoto University, Japan, respectively) have given a condensed review on the U-mart project, which JAFEE ought to commit itself to promote. We then turn to another new frontier. The discipline of econophysics, as described in a previous section, is a new and exciting alternative to financial engineering. This new transdicipline form of statistical physics is not only confined to analyzing financial phenomena, but also embraces a more comprehensive empirical science which R. Nelson has called "process realism." H. Takayasu (Sony Computer Science Laboratories, Japan), the pioneer of econophysics in Japan, gives a concise review of econophysics, which presents the slogan "from microscopic determinism to macroscopic randomness." This approach employs a Langevin-type equation with fluctuating coefficients. Reinforcement mechanisms in economics, which are equivalent to self-organization, are familiar to economists and elucidate an important feature of path-dependency in a Santa Fe context. Here M. Tabata, N. Eshima, and I. Takagi (Kobe University, Oita Medical University, and Kyushu

Tokai University, Japan, respectively), a group of mathematicians, apply the master equation of self-organization in statistical physics to labor mobility due to regional economic disparity.

This study notably uses a new approach to define the transition rate in the master equation by employing the psychological value of individuals or subgroups in line with Helbing (1995) and Feger (1978)[26] and also contains a remarkable numerical analysis requiring about a month to carry out with the use of computer with 1-GHz processor.

3.7 Economic Heresies

Finally, there is a section for a few heterodox economic interpretations which cannot be settled within the familiar classifications of evolutionary economics. A heterodox attempt to reconstruct a realistic economic system with historical time embedded in terms of a neural network is given by M. Yoshida (Senshu University, Japan), whose idea arose from his thorough reading of J.M. Keynes' *Treatise on Money*. On the other hand, in order to thoroughly illuminate the true reality, there is a radical methodological inspection. T. Lawson (University of Cambridge, UK), a keynote speaker at JAFEE2000, stresses that we urgently need a specific project of ontological attention to uncover the nature of being in order to achieve realistic social theorizing. His methodological stand will put almost all of the scientific project in a great peril.

Acknowledgments. We are indebted to the efforts of the copy editors at Springer-Verlag, Tokyo, for improving the language of the manuscripts, of Japanese authors, in particular. As editor of this volume, I express my gratitude to them and to other editorial staff members of Springer-Verlag for their assistance in making this publication possible.

References

Allais M (1953) Le comportement de l'homme rationnel clevant le risque, critique des postulats et axiomes de l'école américaine, Econometrica 21:503–546

Anderson P, Arrow K, Pines D (eds) (1988) The economy as an evolving, complex system. Addison-Wesley, Reading

Arthur B (1994) Increasing returns and path dependence in the economy. University of Michigan Press, Michigan, p 41

Aruka Y (2000) Avatamsaka game structure and its experimental design. Working paper Series 3, Institute of Business Research, Chuo University

Aruka Y (2001) Avatamsaka game experiment as a nonlinear Polya urn process. Proceedings of the JSAI 2001 International Workshop on Agent-Based Approaches in Economic and Social Complex Systems, Matsue City, Shimane, Japan, May 21 and 22, 2001

Feger H (1978) Konflikterleben und Konflikterverhalten. Huber, Bern

---

[26] Helbing (1995) called the psychological value of individuals or subgroups "utility". But his idea, indebted to Feger (1978), is quite different from the usual notion of expected utility neglecting the process reality for a final decision.

Fishburn PC (1982) The foundation of expected utility, D. Reidel, Dordrecht
Frank R (1988) Passions within reason: the strategic role of the emotions. Norton, New York
Frank R, Cook P (1995) The winner-take-all society. Free Press, New York
Harcourt GC (1972) Some Cambridge controversies on the theory of capital. Cambridge University Press, Cambridge
Harcourt GC, Laing WF (1971) Capital and growth. Penguin, Harmondsworth
Helbing D (1995) Quantitative sociodynamics. Kluwer, Dordrecht
Holland J (1975) Adaptation in natural and artificial systems. University of Michigan Press, Ann Arbor
Kagel JH, Roth AE (eds) (1995) Handbook of experimental economics. Princeton University Press, Princeton
Mantegna RN, Stanley HE (2000) An introduction to econophysics. Cambridge University Press, Cambridge
Marschak J (1950) Rational behavior, uncertain prospects and measurable utility, Econometrica 18:111–141
Morishima M (1983) The economics of industrial society. Cambridge University Press, Cambridge
Nelson R, Winter S (1982) An evolutionary theory of economic change. Harvard University Press, Cambridge
Polya G, Eggenberger F (1923) Über die Statistik verketter Vorgange. Z Angew Math Mech 3:s.279–289
Puu T (2000) Attractors, bifurcation and chaos: nonlinear phenomena in economics. Springer, Berlin
Shiozawa Y (1999) Economics and accounting: a comparison between the philosophical background of two disciplines in view of complexity theory. Accounting Auditing Accountability J 12:19–38
Sraffa P (1925) Sulle relazione fra costo e quantità prodotta. Ann Econ 2:277–328
Weiss G (ed) (1998) Multiagent systems: a modern approach to distributed artificial intelligence. MIT Press, Cambridge

Part I Streams of Evolutionary Economics

A Viewpoint on Evolutionary Economic Theory

RICHARD R. NELSON

1 Varieties of Evolutionary Economic Theory

Over the past decade evolutionary economics, or at least bodies of thought and writing that are associated with that term, has become increasingly fashionable. The Schumpeter Society has a strong evolutionary spirit, a society for evolutionary political economy has become well established in Europe, and there have been several summer schools concerned with teaching evolutionary economics. Several years ago, the *Journal of Evolutionary Economics* was launched, and there are several other new journals that clearly have an evolutionary economics flavor. I would like to praise *Industrial and Corporate Change* as an especially interesting journal for economists inclined towards an evolutionary theory. Geoffrey Hodgson has written a book that surveys precursors to modern evolutionary economic theorizing, Ulrich Witt has put together a volume of what he considers is classic work in the field, and now Japan has initiated a society for evolutionary economics.

In this chapter, I will argue that the body of writing that calls itself evolutionary economics is actually quite diverse. However, it is unified by two central features. First, the central interest is in characterizing dynamic processes. In some cases, the principal concern is with economic variables that are understood to be undergoing continuing change. In other cases, the interest is in a particular configuration that is assumed to be in equilibrium, but the explanation for that configuration involves, in an essential way, an analysis of how it got there. That leads me to the second common characteristic of economic writing that call itself evolutionary. Almost all of it assumes, or has built into it, path dependency. Within evolutionary theories, or models, history matters.

However, if one probes beneath these important commonalities, I think one can identify at least three distinguishable subgroups, or schools, of evolutionary theorizing: evolutionary game theory, theorizing that attaches itself to the new mathematical writings on nonlinear dynamic systems, and variation–selection models.

School of International and Public Affairs, Columbia University, 420 West 18th Street, New York, NY 10027, USA

The focus in evolutionary game theory is on repeated games, generally with a given and fixed number of basic strategies available to the actors, and in which there are multiple possible equilibria. Which, if any, equilibrium ultimately is reached is determined by the dynamic "out of equilibrium" evolutionary processes built into the model, and by initial conditions. For the most part, the interest is in how the dynamic processes affect what equilibrium (if any) is ultimately reached. On the other hand, the analysis may purport to explain a particular pattern of behavior, interpreted as an equilibrium in a game, as the result of certain dynamic processes that got the system where it is rather than to another equilibrium point.

The by-now rather extensive body of writing by economists who have developed complex nonlinear dynamic models that purport to represent in abstract form certain classes of economic process seems to me to be of quite a different type. Here, unlike most evolutionary game theory, the focus of attention is on patterns of continuing change. In many cases, simulation is the vehicle for modeling, and the key theoretical arguments and references are to what has come to be understood about the mathematics of nonlinear dynamic systems more generally. Bifurcation points, chaos, and emergent properties are often part of the language used.

While there is some overlap, these two areas of "evolutionary economics" seem to me to be quite different in form and spirit from the kind of evolutionary theory Sidney Winter and I mapped out nearly 15 years ago in our book *An Evolutionary Theory of Economic Change*. Our central argument was that much economic change could be understood as occurring through the operation of a particular kind of dynamic process, and that economic behavior and performance at any time needed to be treated in terms of where that process had led at that time. The dynamic process we described had features analogous with, but details different from, evolutionary processes in biology. Its key features are these. At any particular time, there is usually some variety in the variables of interest. Selection mechanisms operate to reduce the extent that variety, and there are relatively strong inertial forces bearing on what survives selection. However, at the same time there are mechanisms that generate new variety.

We proposed that this kind of theory could explain a wide variety of phenomena of interest to economists. In fact, Winter's original interest in evolutionary economic theory grew out of his scrutiny of the arguments presented by some prominent economists, Milton Friedman and Fritz Machlup in particular, that the standard predictions of neoclassical economic theory about the policies and practices of business firms were the same as those that an evolutionary theory would predict. In a famous article written prior to our joint work, Winter showed that the relationships between the predictions of an explicit evolutionary theory and those of neoclassical theory are very complex, even when one is assuming that the environment within which firms are operating is constant. More generally, he argued convincingly that evolutionary theory needed to be regarded as a distinct alternative to, and not an ancillary support for, neoclassical theory for explaining and predicting phenomena that are the central focus of the latter, such as firm pricing and employment policies. While in many cases the predictions of behavior in equilibrium are similar, the causal mechanisms stressed by the two theories are very different.

My own attachment to an evolutionary perspective came about for different reasons. In my view, and I believe Winter's, the principal value of an evolutionary theory of the sort we developed is that it deals effectively with economic behavior and performance in contexts where change is rapid and unpredictable, and where neoclassical economic theory would appear to be out of its plausible domain. Thus, evolutionary theory deals comfortably with contexts where the assumption that behavior is "in equilibrium" is implausible. My own work has focused on economic growth driven by technological advance, and with the dynamics of competition in industries where technological advance is important, two topics where the equilibrium assumptions of neoclassical theory would seem quite inappropriate. However, there are many other topics of interest to economists where the assumption of equilibrium is awkward at best, such as the consequences of structural change in the economies of eastern Europe.

Since we wrote, there have been a large number of other economists who have worked with, and further developed, this kind of evolutionary theoretical framework. I reviewed this body of work in my recent piece in the *Journal of Economic Literature*, and I now want to focus on two points in particular. One is the key analytical commitments of this branch of evolutionary economics and the sources of its strength, or at least the ones I would stress. The second is the constraints on the areas where evolutionary theory of this sort is likely to be powerful. My belief is that this kind of theory is widely applicable, but there are some phenomena of interest to economists where an evolutionary theory may not provide much illumination.

2 Key Analytical Commitments and Sources of Strength

I noted above that the body of economic theorizing I am focusing upon here has features analogous to evolutionary theory in biology, but the details are different. Speaking for myself at least, I was not drawn to an evolutionary theory of economic change because of the appeal of biological theory, but rather because my observations of processes of economic change strongly indicated that what was going on was a combination of selection on extant variation and the generation of new alternatives. There are a wide variety of models which have these features. Models of biological evolution, and models of economic evolution of the sort we are discussing here, are two distinct subclasses.

Recently, of course, the power of "blind" evolutionary processes to develop life forms and patterns of behavior that are wonderfully sophisticated and well adapted to their environmental contexts has become well recognized. Scholars from all disciplines are familiar with the argument of Richard Dawkin's *Blind Watchmaker* and Daniel Dennett's *Darwin's Dangerous Idea*. Henry Plotkin has touted "Darwin machines" as the principal vehicle through which human knowledge grows, an argument in the spirit of, but somewhat more extreme than, the one Winter and I espouse.

In any case, Darwin's theory of biological evolution met resistance in part because it explained extant variety without recourse to a creator with a "plan." The brand of evolutionary economic theory Winter and I have been advocating has met resistance within our discipline because it aims to explain economic change, and the existing configuration of economic variables at any time, without recourse to the assumption that the economic actors planned, or anticipated, very clearly the full range of consequences of what they were doing.

Unlike the more extreme versions of evolutionary theory about human and organizational behavior, we certainly do not deny that thinking and planning are central aspects of economic behavior. The relationships among belief system, thinking, and behavior are very complex in humans and in human organizations. Most behavior of the sort in which economists are interested is at least "rationalized" if not derived as a consequence of reason.

However, our central premise is that humans are *boundedly* rational, and that learning in the areas in which we are interested has a strong element of trial, error, and try-again behavior. In some of these arenas there is also a strong element of interperson or interfirm competition, but not in all. What is essential in our view of evolutionary economics is that alternative modes of *behavior* compete with each other, and that there are systematic and understandable selection process that tend to winnow out the less adequate. Beyond that, our theoretical position is pragmatic and empirically oriented.

Indeed, another central feature of our brand of evolutionary economic theory is that it is, to a considerable extent, empirically oriented and even *inductive*. When one reads articles that fit within the first two of the economic evolutionary theory camps that I mentioned above—evolutionary game theory and nonlinear dynamic theory—one has the feeling that, for the author, the beauty or the power of the formal analysis is the basic desiderata. Empirical material is brought in mostly as "for instances," i.e., as a means of suggesting that the formal theory might actually correspond to something real out there. I am aware of few instances in which careful discussion of the empirical phenomena came first, and the theorizing was introduced and justified as a way to explain those phenomena.

In contrast, virtually all of the body of evolutionary theorizing of the sort that Winter and I have been doing has been focused on providing a plausible interpretation of certain bodies of empirical data, such as the observed time series of outputs, inputs, and prices that have been generated by *experienced* economic growth. In addition to being strongly oriented toward explaining specific empirical phenomena, our brand of theorizing has also explicitly and self-consciously aimed for what might be called "process realism."

To elaborate what we mean by that, we do not think it satisfying simply to have a model which is capable of generating, and hence "explaining," the basic empirical data under consideration. We think it is important that the processes built into that model should be at least *metaphorically* consistent with what is known about the actual processes being modeled. I mentioned above the particular assumptions about human and organizational behavior built into these kinds of evolutionary models. They are built in because they fit the facts. The details of the modeling of the behavior of the actors in our models, and the modeling of how the actors interact with each other and with their environment, are similarly structured with considerable attention to what is known about the case.

Winter and I feel strongly regarding the merits of this philosophy for doing economic theory. We have couched our argument in terms of the relationships between formal theory and appreciative theory in economics.

We have proposed that most of the solid understanding economists possess is in the form of what we have called "appreciative theory," which is causal analysis that is quite close to the empirical phenomena and policy issues in focus. Appreciative theory represents what the economist studying an issue believes is 'really going on." In contrast, formal theory is more abstract, and laid out with a certain distance from particular bodies of empirical phenomena. Formal theory encompasses only a small portion of what economists know about economic phenomena, and even that portion is a highly stylized version of what is known. This position led us to the argument that, to be really useful in advancing economic understanding, formal theory must be able to engage in dialogue, as it were, with the appreciative theory, which is where most of the economists' understanding resides, and to do that, we argue, the former must be developed to match up with the latter.

The evolutionary economic theory that is the focus of this essay is formal theory in the sense described above. It is an explicit abstract representation of the phenomena under study, rather than simply a way of talking about those phenomena.

However, the *language* of economic evolutionary theory is, in our view, one of its strengths. Evolutionary language seems to be the natural language used by economists to explain many of the phenomena they are studying, from the way businesses respond to a change in demand or supply conditions, to the patterns of long-run economic growth. At least, it is the language often employed by economists when they are not self-consciously trying to theorize. Neoclassical language becomes the standard only when economists are explicitly trying to develop a theoretical argument. A formal evolutionary theory has the great advantage of easily conforming with the economist's informed, but not formal, explanation which we have called appreciative theory.

More generally, evolutionary theory is much more conformable than neoclassical theory with social science thinking outside of economics. Thus, its use enables economists to interact much more effectively, with scholars outside of economics, and this is a significant advantage.

When Winter and I put forward that argument nearly 15 years ago, we had general conformability in mind, and at that time we did not anticipate that relatively formal evolutionary theory, along the lines we were espousing, would become an analytic vehicle widely used by scholars studying a variety of different aspects of cultural and social change. Yet it appears that for scholars not coming from a background economics, an explicit evolutionary theory is a very natural way of understanding change. Indeed, evolutionary theory, along the lines we are discussing here, appears to have been independently discovered by a number of scholars of cultural and social change. The various evolutionary theoretical groups are just now beginning to find each other.

Thus over the last 15 years or so, a sizable community of scholars studying the progress of science has adopted an expressly evolutionary theoretical perspective on the processes that are involved. Indeed, Donald Campbell, one of the pioneers in this arena, coined the term "evolutionary epistemology," which is basically a commitment to the proposition that knowledge evolves. Campbell was originally a social psychologist. Most of the modern evolutionary epistemologists have their training outside economics.

Technological historians and sociologists studying technological change, as well as economists, now are in wide accord that the processes of technological advance need to be understood as "evolutionary." This discovery, or the development of this particular theoretical perspective on technological change, was arrived at more or less independently by the different disciplinary camps.

Alfred Chandler, who trained as a historian and who carries out his research in a historian's style, has clearly seen change in standard modes of business organization as proceeding through an evolutionary process. A new generation of business historians and organization theorists are following that theoretical road.

The fact that economists and noneconomists studying these phenomena use the same language, and have in mind a similar form of theory, allows knowledge to be combined. As I noted, this is beginning to happen.

A number of scholars from economics and various other fields have proposed that the law evolves. While in the early stages of the renaissance of interest by economists in institutions neoclassical theory was the basis of the theorizing, in recent years economists interested in institutional change have joined sociologists and historians in using the language of evolutionary theory. However, I believe that these strands of evolutionary theorizing have met with less success than evolutionary analysis of technological change and of change in business organization. It may be useful to consider way.

3 Limitations of Evolutionary Economic Theory

Evolutionary theory is proving to be a flexible analytic instrument. Suitably tailored to the subject matter under study, it has proved to be capable of providing a convincing story about the processes of change governing a variety of phenomena of interest to the broader community of social scientists, as well as to economists. The theory has been effective both when the focus is on explaining an existing state of affairs, and when the orientation has been towards explaining patterns of continuing change.

The power of an evolutionary theory depends strongly on the analyst's ability to specify, in a convincing way, the nature of operative selection criteria and mechanisms. Most extant evolutionary models developed by economists give explicit attention both to a pattern of behavior or type of capability, e.g., a pricing policy or a technology for producing something, and to the economic units that exhibit that behavior or operate that capability. For the most part those units—usually presumed to be business firms—are treated as if they were in competition with each other for customers in a given market. Almost always in the formal models, the criterion of fitness is "profit." This formulation gives rise to a class of powerful models in which both behavior or capabilities and firms are selected in terms of their profitability. Of course probing deeper requires analysis of just what the market treats as profitable.

I noted above, however, that this analytic structure, which involves units as well as practices in competition with each other, is not an essential aspect of evolutionary theories of cultural change. What makes the theory work is its ability to specify a plausible criterion of merit, and a selection mechanism that selects for that. Market competition provides one way of specifying an evolutionary theory, but in many contexts and arenas where one is tempted to apply evolutionary theory there is no real operative market, only (maybe) a metaphorical one. Thus, there is no real "market" that evaluates new science or new law, and even where there is an operative market, evolutionary analysis need not model that market explicitly.

It is interesting that, while in evolutionary theories of technological change developed by economists there are generally firms that use extant technologies and create new ones, in the evolutionary theories of technological change laid out by historians or sociologists, the firms are often ignored, and the competition is seen simply as among the technologies. A similar difference obtains between writing by economists and that by historians and other social scientists regarding the evolution of business form and practice. In most theories that propose that scientific knowledge evolves, while it is recognized that scientists compete for acclaim, the focus is on the competition among theories.

To date, economists and other social scientists have been relatively successful with evolutionary theory when the analyst can persuasively argue either that there is a real "market" operative in the arena, and also be able to specify what the market wants, or that there are generally accepted and enforced criteria for what is better than what (such as, for example, an engineering or scientific consensus). The difficulties which evolutionary theorists have been having with the law and with institutions reflects, in my view, that the selection criteria and mechanisms are extremely complex, and undoubtedly are not constant over time. If that is the case, there may not be much that an evolutionary theory can illuminate.

The issue here is partly concerned with what one includes and excludes under the broad rubric of evolutionary theory, and what kind of illumination one regards as valuable. Thus, women's and men's clothes fashions change over time. These changes are pulled by changes in demand, but changes in demand are also pushed by people's perceptions of what other people are buying. Eric Abrahamson has argued that many kinds of business practices need to be understood as reflecting 'fashions" whose legitimacy is strongly influenced by the fact that others have adopted those practices. *If we choose*, we can say that fashions in clothes and business practices evolve. However, the understanding lent here by evolutionary language and theory is not the power to predict, or even to explain in terms of stable systematic selection forces. Rather, if the theory is basically correct, the understanding is that these phenomena are shaped to a significant extent by random forces, and that the attempt to explain them in terms of "efficiency" may be a fool's quest.

The conceptual and analytic issues here are not simple for someone who espouses an evolutionary theory of economic behavior and performance. In particular, they focus attention on a particular aspect of much economic evolutionary theorizing—the presumption that the practices that one observes are "functional" in some sense for the entities that have adopted them. Many years ago, John Elster questioned the use of "functional" theories in the social sciences, while at the same time granting the Nelson–Winter theory approval as an exception. I noted above that most evolutionary theorizing by economists links the adoption of a practice by an entity to its profitability, which in turn is linked to its ability to survive. This theoretical strategy both allows a functional argument to go through, and eliminates the need to take explicit account of belief systems and conscious goal-directed decision making.

Even so, we still need to consider aspects of human individual and organizational practice where survival is not at stake. It would appear then that an evolutionary theory needs to deal explicitly with belief systems. Science is one case in point, but clothes fashions and some aspects of business practice may be others.

As the above remarks suggest, I regard the ability to specify selection criteria and mechanisms as essential to effective evolutionary modeling. In many cases, it also seems essential to be able to specify in some detail the mechanisms that generate innovation. Here, evolutionary economic theory is still rather backward.

To date, for the most part, evolutionary economic theories have assumed either that new variety (innovation) is a marginal change from extant variety, or that the process is strictly random, or some of both. This has sufficed for the creation of formal models of technological and economic change that have aimed to explain relatively broad stylized facts. However, clearly they will not suffice if the focus is on more fine-grained phenomena such as the development of computer technology, and changes over time in the structure of that industry.

In fact, Winter and I, along with Franco Malerba and Luigi Orsenigo, are now working on this specific problem, and we are calling the new kinds of evolutionary models we are developing "history-friendly."

How Can Evolutionary Economics Evolve?

Geoffrey M. Hodgson

> We shall not cease from exploration
> And the end of all our exploring
> Will be to arrive where we started
> And know the place for the first time.
>
> T. S. Eliot, *Little Gidding*

Summary. This chapter addresses the current growth and popularity of "evolutionary economics" and examines its meaning and potential. The growth in "evolutionary" ideas in economics is put in historic context before examining the various strands of evolutionary theory. The link between modern evolutionary economics and the "old" institutionalism of Veblen and others is explored, alongside the Marxian, Austrian, and Schumpeterian traditions. Returning to the modern situation, it is suggested that formalist strains of both evolutionary and mainstream economics have difficulty in incorporating the key evolutionary concept of novelty. Examining both the historic legacy and the philosophical underpinnings suggests a distinctive identity based on the incorporation of emergent properties and the rejection of reductionist theorizing. The development of modern evolutionary economics may be all the stronger in doing so if it acquires a clear understanding of its own philosophical underpinnings and its own past.

Key words. Evolutionary economics, Formalism, Reductionism, Novelty, Emergence

The growth in interest in "evolutionary economics" is one of the major international academic events of the 1990s. But what is the basis of this success? And will this movement endure well into the twenty-first century, or prove to be a passing fad? In this chapter I shall consider the nature of the current interest in evolutionary economics. This involves a discussion of both the meaning of the term and the intellectual context in which it has been formed and developed. Remarkably, the term is highly ambiguous. The concern must be that the failure to establish a distinctive identity will lead to the fusion of evolutionary ideas with the limited and overly formalistic approaches of mainstream economics.

The Business School, University of Hertfordshire, Mangrove Road, Hertford, Hertfordshire SG13 8QF, UK

The basis for a broad but distinctive identity for evolutionary economics is sketched out here in the hope that it will be recognized and adopted by its exponents. It is argued that to improve the chances of both survival and development, evolutionary economists must be conscious both of their predecessors in the early twentieth century, and of the philosophical underpinnings of a distinctive evolutionary approach.

1 The Fashionable Nature of the Term "Evolutionary" in Economics

In economics today, the term "evolutionary" is highly fashionable, but this is a relatively recent phenomenon. From the 1920s to the 1960s, the use by economists of biological analogies or evolutionary terminology was quite rare. Notable exceptions existed, such as Armen Alchian's (1950) famous article, and the use of the word "evolutionary" by Joseph Schumpeter.

There was another period, before the First World War, when evolutionary phraseology and biological analogies were prominent in economics and in the social sciences generally. This period saw both the general rise of "social Darwinism" and, in particular, the global popularity of the writings of the English evolutionary polymath Herbert Spencer. The German historical school had made strong use of organic and biological analogies in economics (Hutter 1994). The American neoclassical economist John Bates Clark laced his *Philosophy of Wealth* (1885) with organic metaphors and images taken from Spencerian biology. Likewise, Spencerian theoretical ideas pervade Alfred Marshall's seminal *Principles of Economics* (1890). Marshall openly acknowledged Spencer as a supreme influence on his thought (Hodgson 1993).

In this context, the evolutionary economics of Thorstein Veblen and his "institutionalist" colleagues emerged in the United States. Veblen (1898, 1899, 1919) made extensive use of Darwinian ideas in his attempt to transform the foundations of economic science. However, by the First World War, the use of evolutionary ideas and biological analogies were becoming widely unpopular in Western academia (Degler 1991). The tide of opinion against evolutionary ideas was such that even Schumpeter (1934 p. 57) accepted, in 1911, that "the evolutionary idea is now discredited in our field." This not only created problems for the American institutionalists. The entire direction of social science was altered.

Evolutionary thinking and biological analogies reappeared in Western social science after the Second World War. The use of word "evolution" became quite common in anthropology in the 1960s, although it remained controversial. Some years later it began to be used by economists, many of whom were seemingly ignorant of the precedents before the First World War. The numbers of relevant works in English in economics encountered by the present author in the years from the 1890s to the 1980s are shown in Fig. 1.[1] It has to be born in mind that the overall number of works published in economics and other sciences has increased exponentially in the twentieth century as a whole, and explosively since 1945. Taking this into account, there

[1] In Fig. 1, subsequent editions of the same book are ignored.

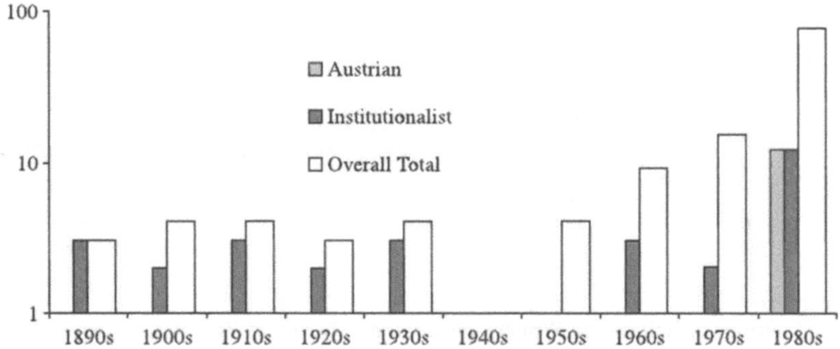

FIG. 1. Number of works in economics (in English) encountered from the 1890s to the 1980s with the word "evolution" or derivatives in the title or subtitle. (logarithmic scale on the vertical axis)

was a relatively significant usage of the term evolutionary in the 20 years prior to 1914. Much of this early usage of the term was by institutionalists of the old school. The number published in this category since 1990 is well into the hundreds.[2]

If we used a linear vertical scale this would give an impression of an explosive growth in evolutionary economics from the 1980s. The earlier period of prominence of the term—from the 1890s to the 1930s—would be somewhat dwarfed by the 1980s. We compensate for this perceptive bias, and the general explosion of the overall number of publications in economics in the last three decades of the twentieth century, by showing the data in Fig. 1 on a vertical logarithmic scale.[3]

Figure 1 underlines the popularity of evolutionary thinking from the 1890s to the 1930s, and the dominance of works in that area by institutionalist authors.

Let us turn again to the 1980s. To a large degree, the increasing use of the term evolutionary economics today can be traced to the impact of Richard Nelson and Sidney Winter's classic (1982) work *An Evolutionary Theory of Economic Change*, although other developments in both orthodox and heterodox economics in the 1960s, 1970s, and 1980s are also important. Previously largely confined to the institutionalist, Austrian and Schumpeterian fringes, the word evolutionary did not become widespread in postwar economics until after 1982.[4]

The upward trend of citations of Nelson and Winter's book is also indicative of the growing influence of evolutionary ideas. This growth is shown in Fig. 2. In the last 30

[2] The use of the word "evolution" does not necessarily indicate the adoption of a biological metaphor. Furthermore, the "evolution-in-the-title-or-subtitle" criterion is very crude. Note that the use of the word evolution in titles by Austrian authors appeared earlier, but did not become common until the late 1980s. Furthermore, Schumpeter did not use the term evolution in a title of any book or article. Also, the term "neo-Schumpeterian" did not become prominent until the 1980s. However, the term evolution was relatively prominent in institutionalist writings from the 1890s until the 1930s.

[3] In this case, the values zero and unity are both represented by a column of zero height.

[4] But note also Axelrod and Hamilton (1981), Boulding (1978, 1981), Nelson and Winter (1973, 1974), and the advocacy of closer links with biological and other sciences by Georgescu-Roegen (1971).

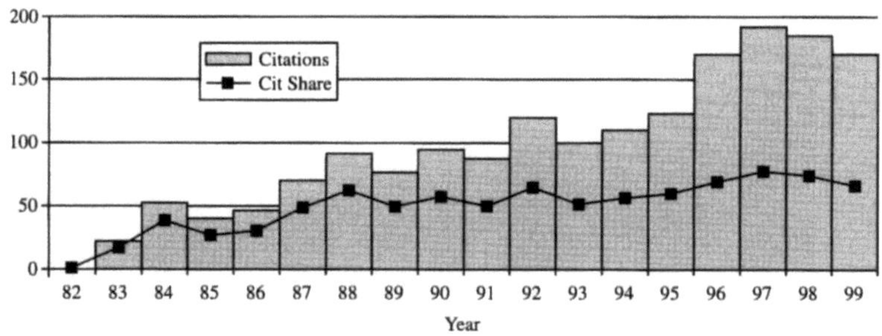

FIG. 2. Citations to Nelson and Winter's *Evolutionary Theory of Economic Change*

years, the overall number of citations in social science has increased more than tenfold. The explosion in the overall volume of all citations has to be taken into account in time-series citation analysis. We may divide the actual number of citations for the work in question for a given year by the total volume of citations in the social sciences. This gives the "citation share" (per million) in Fig. 2.

By the late 1980s, work in evolutionary economics had been broadened and accelerated by the growth in both America and Europe of various institutional, Austrian, and Schumpeterian approaches to economics.[5] There were notable and fruitful applications of these ideas, particularly in the sphere of technological change (Dosi et al. 1988; Freeman 1990; Hanusch 1988). Evolutionary economics established an impressive research programme and had a major impact on economic policy, particularly in the areas of technology policy, corporate strategy, and national systems of innovation. A substantial body of work is now clearly visible, and worthy of reflective evaluation.

Nevertheless, there is still no established consensus on what evolutionary economics should mean. Many economists use the term whilst wrongly taking it for granted that a common and obvious meaning is implied. Indeed, the use of the word evolutionary in economics seems to be very much a matter of vogue. It seems fashionable to publish any work evoking any kind of dynamic consideration with the term evolutionary in its title. Even—or especially—in the context of widespread popularity, this is an unsatisfactory, and possibly hazardous, state of affairs. We now move on to consider the possible meanings of the term.

2 Meanings of "Evolutionary Economics"

The term "evolutionary economics" is currently applied to a confusingly wide variety of approaches within the subject. At least six main groupings using the phrase can be identified.

[5] The European Association for Evolutionary Political Economy and the International Joseph Schumpeter Association were both formed in the late 1980s. The *Journal of Evolutionary Economics* commenced publication in 1991, and articles on evolutionary economics have also appeared in the *Journal of Economic Issues* (published by the US-based Association for Evolutionary Economics), the *Journal of Economic Behavior and Organization*, and other journals.

— A century ago Veblen (1898, 1919) argued for an "evolutionary" and "post-Darwinian" economics. Institutionalists in the tradition of Veblen and John Commons frequently describe their approach as evolutionary economics, often using the terms "institutional" and "evolutionary" as virtual synonyms, as exemplified in the title of the Association for Evolutionary Economics—the USA-based association of institutional economists.
— Schumpeter (1942) famously described capitalist development as an "evolutionary process." Work influenced by Schumpeter is also described as evolutionary economics, as evidenced by the title of the *Journal of Evolutionary Economics*, published by the International Joseph Schumpeter Association.
— The approach of the Austrian school of economists is often described as evolutionary, as portrayed in Carl Menger's theory of the evolution of money and other institutions, and by the extensive use of an evolutionary metaphor from biology in the later works of Hayek, especially in relation to the concept of spontaneous order.
— In addition, the economics of assorted writers such as Adam Smith, Karl Marx, Marshall, and others is also sometimes described as evolutionary in character.
— Evolutionary game theory is a prominent recent development in mathematical economics and has been inspired by related mathematical work in theoretical biology.
— The word evolutionary is sometimes attached to work in what is also described as "complexity theory," typically that associated with the Santa Fe Institute in the United States, involving applications of chaos theory and various other types of computer simulation. In this and allied simulation work, the use of replicator dynamics, genetic algorithms, genetic programming, and so on can be found.

Nothing is more guaranteed to generate confusion and to stultify intellectual progress than to raise a muddled term to the centerpiece of economic research, whilst simultaneously suggesting that a clear and well-defined approach to scientific enquiry is implied. It is important both to sort out the different meanings of the term and to consider carefully its conceptual history.

3 Taxonomizing "Evolutionary Economics"

In previous work (Hodgson 1993, 1997), I have proposed two taxonomies of relevant meanings of "evolutionary economics." These taxonomies are ways of seeing distinctions between different types of approach. They are complementary rather than contradictory, because each focuses on different aspects of the problem. These two taxonomies are briefly summarized here. The object is to identify key features of various approaches which are relevant to the survival and development of evolutionary economics within the discipline.

The taxonomy in Hodgson (1993) deals with differences between the mechanisms of evolutionary change proposed by different authors. A major focus is on the important difference between "ontogenetic" and "phylogenetic" conceptions of change. This distinction is useful because it exposes the more limited character of the former type of evolution. Note that use of these terms should not be taken to imply an endorsement of the view that human behavior is determined by genes.

In biology, ontogeny involves the development of a particular organism from a set of given and unchanging genes. Its environment will also affect its development, but nevertheless the growth of the organism is the result of genetic instructions. Hence, the genes represent a given set of (environmentally dependent) developmental possibilities. In contrast, phylogeny is the complete and ongoing evolution of a population, including changes in its composition and that of the gene pool. It involves changes in the genetic potentialities of the population, as well as their individual phenotypic development.

By analogy, in economic evolution ontogeny traces institutional and other developments in the context of an environment but with fixed "genetic material." If we reject the argument that socioeconomic evolution can be explained in terms of the human biotic inheritance, then an alternative and analogous supposition is to assume inert individuals with given purposes or preference functions. For example, for the purposes of theoretical explanation, given individuals are assumed in Menger's (1981) account—first published in 1871—of the evolution of money, and Hayek's (1982, 1988) discussion of the evolution of spontaneous order. By contrast, Veblen (1898, 1899, 1919) gives more emphasis to changing purposes, preferences, habits, and beliefs in his accounts of economic evolution. Indeed, his major criticism of other economists—from Menger to Marshall—was that they typically adopted "a faulty conception of human nature." As Veblen (1898 p. 389) elaborated:

> In all the received formulations of economic theory, whether at the hands of English economists or those of the Continent, the human material with which the inquiry is concerned is conceived in hedonistic terms; that is to say, in terms of a passive and substantially inert and immutably given human nature.

Veblen's advocacy of a "post-Darwinian" approach to economics involved a full, phylogenetic conception of change, in which human preferences as well as social institutions evolved.

However, as in biology, phylogenesis subsumes ontogenesis. The phylogenetic development of a population includes the ontogenetic development of the individuals within it. Hence the objection here to Menger's account of the evolution of money, or Hayek's description of the evolution of spontaneous order, is not so much that they are wrong, but that they are only part of the story. Indeed, in some other passages in his later work, Hayek (1982, 1988) goes some way to broadening the evolutionary picture and considering changes in cultures and individual habits. Accordingly, the distinction between ontogenetic and phylogenetic conceptions of evolution does not involve mutually exclusive categories. However, approaches that are confined to ontogeny are not as broad and developmental as the full, phylogenetic approach. This will be referred to below as the "genetic criterion," but again with a warning that the word genetic does not refer to the determination or otherwise of behavior by genes. The distinction between ontogeny and phylogeny concerns the broad mechanisms and outcomes of evolutionary change.

In Hodgson (1997), an alternative classification was attempted, focusing on fundamental philosophical issues. This pays particular attention to the varied ontological and methodological foundations of the theories involved. One of the differences discussed was over the use, or otherwise, of biological metaphors in evolutionary economics. Those that have argued that various biological analogies are useful in economics include Marshall (1890), Veblen (1919), Georgescu-Roegen (1971), Nelson and Winter (1982), and Hodgson (1993). Others have distanced themselves in varying degrees from biological metaphors (Schumpeter 1954; Witt 1992; Foster 1997). However, in economics, as in science general, overt stances concerning metaphor do not always correspond with the hidden uses and roles of metaphor in a theory. Accordingly, proclamations concerning the value or otherwise of biological metaphors often reveal more about philosophical prejudices than about what is really going in the theory. As a result, the use or otherwise of biological metaphors is not an entirely reliable or clear-cut taxonomic criterion.

For these and other reasons, two criteria were given preeminence in the second taxonomy (Hodgson 1997). The one given the most stress was the ontological criterion of novelty. This was followed by the methodological criterion of reductionism. Arguably, considerations of ontology are the more fundamental. We shall discuss the methodological criterion first.

The methodological criterion concerns whether explanations in evolutionary economics are reductionist or nonreductionist. Reductionism sometimes involves the notion that wholes must be explained entirely in terms of their elemental, constituent parts. More generally, reductionism can be defined as the idea that all aspects of a complex phenomenon must be explained in terms of one level, or type, of unit. Explaining individuals solely in terms of social structures, i.e., the culture of institutions, would be an example of reductionism. According to this view, there are no autonomous levels of analysis other than this elemental foundation, and no such thing as emergent properties upon which different levels of analysis can be based.

The issue of reductionism is an important one in modern economics, but it divides evolutionary economists. In social science in the 1870–1920 period, reductionism was prominent and typically took a biological form. Accordingly, attempts were made to explain the behavior of individuals and groups in terms of their alleged biological characteristics. Reductionism is still conspicuous in social science today, and typically appears as methodological individualism. This is defined as "the doctrine that all social phenomena (their structure and their change) are in principle explicable only in terms of individuals—their properties, goals, and beliefs" (Elster 1982 p. 453). It is thus alleged that explanations of socioeconomic phenomena must be reduced to properties of the constituent individuals and relations between them.[6]

Allied to this is the sustained attempt since the 1960s to found macroeconomics on "sound microfoundations," as critically reviewed by S. Abu Turab Rizvi (1994). There

[6] Note that this popular term is sometimes used in additional, ambiguous, and contradictory ways. Confusingly, Winter (1988) defends "methodological individualism" while elsewhere invoking concepts such as organizational knowledge and group learning (Winter 1982). However, what Winter seems to have in mind here by methodological individualism is a rejection of the idea that intentions and interests can be attributed to groups and organisations, rather than to individuals. Contrary to Winter, the emphasis here is rightfully on the issue of explanation, not on the attribution or nonattribution of qualities to individuals or groups. The idea that intentions and interests should not be attributed, at least in an unqualified manner, to groups and organizations equally as well as to individuals is compatible with a rejection of methodological individualism. It is arguable that in the proper, methodological (i.e., explanatory) sense, Winter is not a methodological individualist.

are other versions of reductionism, however, including versions of "holism" that suggest that parts should be explained in terms of wholes. Reductionism is countered by the notion that complex systems display emergent properties at different levels that cannot be completely reduced to, or explained wholly in terms of, another level. By contrast, antireductionism generally emphasizes emergent properties at higher levels of analysis that cannot be reduced to constituent elements. Accordingly, notions such as group selection in biology and group knowledge in the social sciences can be sustained. Antireductionism in the social and biological sciences is crucially associated with the concept of emergent properties (Archer 1995; Bhaskar 1979; Hodgson 1988, 1993; Kontopoulos 1993; Lawson 1997; Mayr 1985; Murphy 1994).

The ontological criterion concerns whether or not substantial emphasis is given to the assumption that evolutionary processes in economics involve ongoing or periodic novelty and creativity, thus generating and maintaining a variety of institutions, rules, commodities, and technologies.

Conceptions of economic evolution that stress novelty typically highlight indeterminacy and the possibility of cumulative divergence, in contrast with convergence and equilibria (Andersen 1994; Foss 1994; Hodgson 1993; Witt 1987). Notably, the Austrian school of economists give pronounced stress both to the indeterminacy and the potential novelty of human imagination, action, and choice (Lachmann 1977; Loasby 1976; Shackle 1955). Kenneth Boulding (1991 p. 13) wrote: "One very fundamental principle in evolutionary processes is their profound indeterminacy." Outside economics, Karl Popper's stress on indeterminacy, novelty, and emergent properties should be mentioned (Popper 1982). However, novelty does not necessarily involve indeterminacy. For instance, chaos theory highlights potential novelty and divergence, and does this using unpredictable but essentially deterministic systems.

These three binary criteria—ontological, genetic, methodological—give eight possible classifications, as shown in Fig. 3. The ontological is treated as the most fundamental, given the importance of the generation of novelty in systems which reasonably could be described as evolutionary. To save space, I shall not spend time justifying the allocation of individual theorists, save to add a few caveats. First, taxonomies are always imperfect: rich and complex thinkers are often difficult to classify. Second, in some cases, notably Hayek, there is a shift of view from one category to another. Here, for instance, "Hayek I" refers to his thought up to the 1960s, to be followed by "Hayek II," where a richer conception of socioeconomic evolution is developed.[7]

Note, however, that two notable subsets formed by (i) theorists who embrace novelty and ontogenetic conceptions of change, and (ii) theorists who fail to stress novelty and are disposed to phylogenetic conceptions of change, are both empty. This does not imply that the ontological and genetic criteria are identical. Indeed, there may be theorists who fit in these subsets. What it does show, however, is a strong connection between those who assume novelty in evolutionary processes and those who

[7] Compared with Hodgson (1997), I have changed the classification of Witt in this schema. On closer inspection, although he acknowledges the role of groups, Witt does not seem to have an elaborated notion of emergent properties that could sustain a developed antireductionist stance. In general, taxonomies involve difficulties of interpretation and classification, and this is no exception.

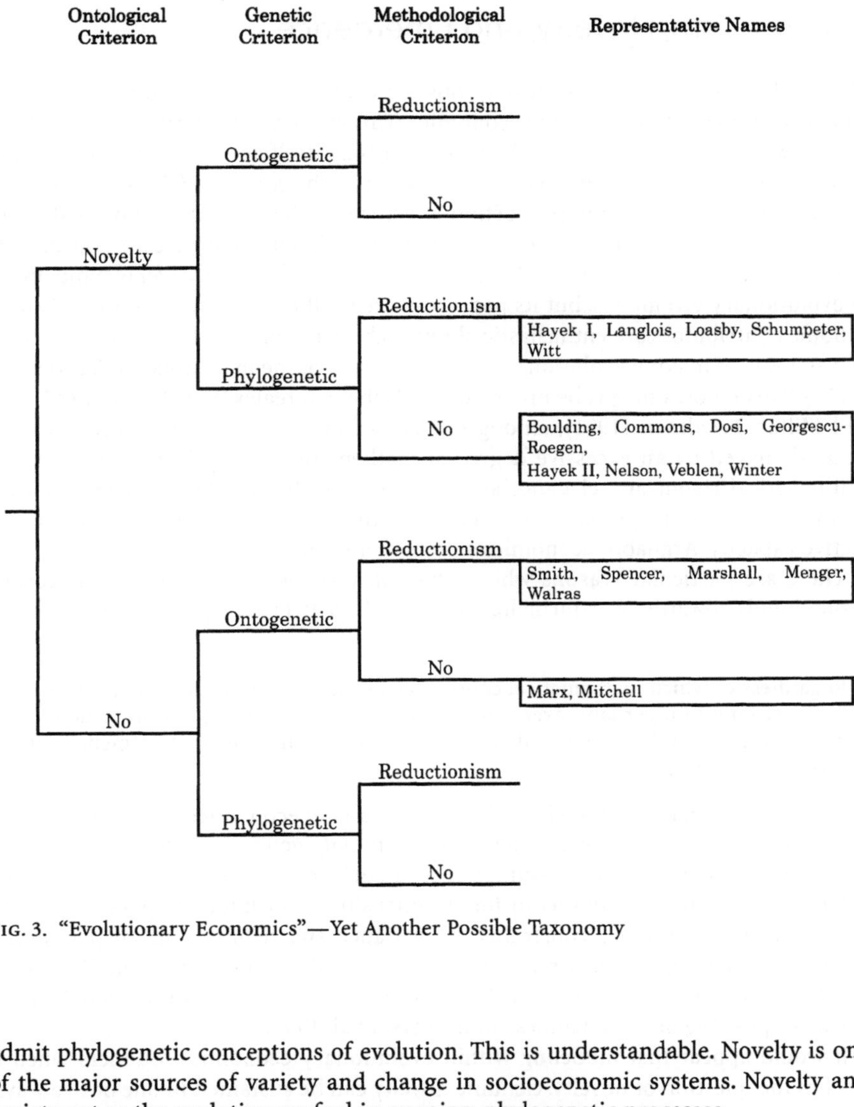

FIG. 3. "Evolutionary Economics"—Yet Another Possible Taxonomy

admit phylogenetic conceptions of evolution. This is understandable. Novelty is one of the major sources of variety and change in socioeconomic systems. Novelty and variety act as the evolutionary fuel in ongoing, phylogenetic processes.

We should also note the concentration of theorists in the novelty-embracing, phylogenetic, and antireductionist category. This is not only where much of the innovative action is in modern evolutionary economics (Dosi, Nelson, Winter), but also a category that includes earlier institutional and evolutionary thinkers such as Veblen and Commons. Apart from the question of reductionism, they are close to an other important group of evolutionary economists influenced by the later Hayek (Langlois, Loasby, Witt). I have argued recently that this type of approach is potentially the most fruitful from the point of view of evolutionary economics (Hodgson 1993, 1997, 1999).

4 Formalism, Novelty, and Emergence

The fundamental question of novelty raises a central issue concerning the survival and development of evolutionary economics. This concerns the limits to formal modeling. It should be noted that much work describing itself as evolutionary economics makes extensive use of mathematics. For example, the *Journal of Evolutionary Economics* has published a high proportion of articles with a high degree of mathematical formalism, comparable with the most prestigious mainstream journals in economic theory.[8] The *Journal of Economic Behavior and Organization* is sympathetic to evolutionary economics, but its pages are often full of mathematical formalism.

Some contributors are enthusiastic about such mathematical developments, seeing this as a way of making evolutionary economics part of the mainstream (Heertje 1994 p. 275). This ignores the problem—raised by Nobel Laureates Wassily Leontief (1982) and Milton Friedman (1991), among others—of the extent to which economics is being damaged by an excessive degree of mathematical formalism. In our subject, mathematical rigour and elegance are now rated much more highly than empirical richness, explanatory power, conceptual robustness, policy usefulness, or even predictive capacity. Arguably, economics suffers as a result.

There are particular reasons why evolutionary economists should be concerned about trends towards over-formalization. Ulrich Witt (1992 pp. 9–10) hints at this when he writes:

> Regardless of which mathematical concept seems the most promising, none of them have yet found or, perhaps, even considered an answer to the question of how the emergence of novelty may be adequately represented in a formal treatment of the evolutionary process.

Arguably, mathematical modeling taken to its limits in economics would greatly constrain novelty. Even if chaotic systems can seemingly generate novelty, they are still limited by their own formal assumptions. In such models, the system needs to be defined by formal assumptions in finite dimensions that limit the possible results. Even a stochastic process constrains the variance and defines a given parametric space. By its nature, novelty strains against the boundaries of formalism. To endogenize the novelty-creating process within a formal framework is always greatly to limit the set of possible novel outcomes (Bonaccorsi et al. 1995).

It would appear that modeling within evolutionary economics founders when it reaches the question of novelty. Genuine novelty can be sustained in one of two ways. One is to postulate the possibility of an "uncaused cause." Hence, George Shackle (1989 p. 51) asserts that "economics is about choice as a *first cause*, that is the coming into being of decisive thoughts not in all respects to be explained by antecedents." Ludwig Lachmann (1969 p. 93) comes from the same intellectual tradition, arguing that spontaneous mental action is "not a 'response' to anything pre-existent," and thus it is wholly spontaneous and undetermined. Their common antideterminism is based on a conception of the essential indeterminacy of human decision-making: of individual

---

[8] However, current *JEE* editorial policy is to encourage nonmathematical as well as mathematical articles.

decision and action as a first or uncaused cause. The problem with this argument is that it rests uneasily with the attempts of science to find causal explanations. It is also nonevolutionary, at least in the sense that it cannot give the emergence of the uncaused will in higher animals an evolutionary explanation.

An alternative is to establish the concept of emergence. It is to see consciousness, purposefulness, and will as *emergent properties* of a highly developed mental apparatus. As Tony Lawson (1997 p. 176) explains: "an entity or aspect is said to be *emergent* if there is a sense in which it has arisen out of some 'lower' level, being conditioned by and dependent upon, but not predictable from, the properties found at the lower level." The concept of emergence has played an important role in science for over 100 years. It makes it possible to retain causality, in the sense that human intentions are caused by something, but also to treat them *as if* they were uncaused.

From this perspective, the concept of emergent properties plays a crucial role. The concept makes it possible to have the best of both worlds: we can have causality and novelty at the same time. As Leonard T. Hobhouse (1913), William McDougall (1929), and others suggested long ago, concepts such as consciousness and purposeful behavior can be regarded as emergent properties of highly complex nervous systems in highly evolved organisms. The concept of emergence makes it possible to retain causality, in the sense that human intentions are caused by something, but also to treat intention or will as an emergent property. Hence intention *appears* as a different type of cause, although ultimately it depends completely on materialist causal interactions within the human nervous system.

5 To Arrive Where We Started

Strikingly, this agenda—combining the philosophical concepts of emergence and antireductionism with the broader evolutionary concept of phylogenetic change—brings us back to the unfinished business of the institutionalism of the 1890s and the early twentieth century.[9] In this section we will outline the earlier parallels, before moving on to related modern developments in the next section.

Although he did not complete the argument, Veblen moved toward a reconciliation of potentially novel and purposeful behavior with the notion of an efficient cause. Writing in 1898, Veblen (1934 pp. 80–5) wrote:

> By selective necessity he is endowed with a proclivity for purposeful action. . . . He acts under the guidance of propensities which have been imposed upon him by the process of selection to which he owes his differentiation from other species.

The human attribute of purposefulness was thus a result of evolution itself. Accordingly, the mainstream assumption of rationality lacked an account of its own evolutionary emergence. In other words, would a rational "economic man" emerge in

[9] Both authors have now rectified the omission of any discussion of Veblen and the "old" institutionalism from Nelson and Winter (1982). Winter (1990) has referred approvingly to Veblen. Nelson (1995) has recognised the evolutionary contribution of Veblen and gone so far as to write: "[I now] understand fully that I am part of the old institutional economics" (Nelson, personal communication, 1996).

human evolution? Again writing in 1898, Veblen (1934 p. 79) observed the lack of an adequate answer to this question, and an insufficient realization by mainstream economists that the answer may be negative:

> But if this economic man is to serve as a lay figure upon which to fit the garment of economic doctrines, it is incumbent upon the science to explain what are his limitations and how he has achieved his emancipation from the law of natural selection.

Consider the related concept of emergence. This concept emanates originally from the British philosopher C. Lloyd Morgan (1896, 1923, 1927). In another work, I have argued that a crucial visit of Morgan to Chicago in 1896 was inspirational for Veblen and led to his crucial concept of institutional evolution (Hodgson 1998). The idea of emergent properties gave Veblen an argument against the reduction of economics to *biological* foundations, and against the fashionable biological determinism of the 1890s. Veblen stated that if socioeconomic phenomena were determined exclusively by biological factors, then the concepts of institution and culture would be redundant. In his view, culture and institutions were irreducible to biological factors alone. Although he did not use the phrase "emergent property" explicitly, this tied in with the view that culture and institutions, although they depend on the biotic foundation, have emergent properties that are irreducible to biotic components. Accordingly, the concepts of cultural and institutional evolution were developed in *The Theory of the Leisure Class*. However, Veblen's theory was incomplete, and much more work is still required to develop an adequate and systemic theory of institutional evolution.

Covertly, the concept of emergence played another role in the development of institutionalism. By reliance on the concept of emergence, the relative autonomy of macroeconomics and the idea of the workability of aggregates can be established. In his 1924 Presidential Address to the American Economic Association, the institutional economist Wesley Mitchell (1937 p. 26) argued that economists need not begin with a theory of individual behavior, but with the statistical observation of "mass phenomena." Mitchell (1937 p. 30) went on: "The quantitative workers will have a special predilection for institutional problems, because institutions standardize behavior, and thereby facilitate statistical procedure." Subsequently, Rutledge Vining (1949 p. 85) noted how "much orderliness and regularity apparently only becomes evident when large aggregates are observed," and noted the limitations of a reductionist method in economics. Modern computer simulations and other studies of complex systems seem to underline similar points (Cohen and Stewart 1994; Chiaromonte and Dosi 1993).

Mitchell and his colleagues in the US National Bureau for Economic Research in the 1920s and 1930s played a vital role in the development of national income accounting, and suggested that aggregate, macroeconomic phenomena have an ontological and empirical legitimacy. Arguably, this important incursion against reductionism in economics created space for the Keynesian revolution. Through the development of national income accounting, the work of Mitchell and his colleagues helped to establish modern macroeconomics, and influenced and inspired the macroeconomics of Keynes (Mirowski 1989 p. 307).

The early institutional economics did not attempt to build up a picture of the whole system by moving unidirectionally from given individuals. Instead, there was the idea of interactive agents, mutually entwined in durable and self-reinforcing institutions.

This suggested a quite different way of approaching the problem of theorizing the relationship between actor and structure.

The "old" institutionalism saw institutions as connected to individual habits. Indeed, an institution was defined by institutionalists in the old tradition as "a way of thought or action of some prevalence and permanence, which is embedded in the habits of a group or the customs of a people.... Institutions fix the confines of and impose form upon the activities of human beings" (Hamilton 1932 p. 84). Habits both reinforce and are reinforced by institutions. Through this circle of mutual engagement, institutions are endowed with a stable and inert quality, and tend to sustain, and thus "pass on," their important characteristics through time. Further, institutions play an essential role in providing a cognitive framework for interpreting sense-data, and also intellectual habits or routines for transforming information into useful knowledge. The strong influence of institutions upon individual cognition provides some significant stability in socioeconomic systems, partly by buffering and constraining the diverse and variable actions of many agents.

A rigorous and detailed exposition is still lacking, but we may sketch out a possible argument along the following lines. The institutional*izing* function of institutions means that macroeconomic order and relative stability are reinforced alongside variety and diversity at the microeconomic level. Ironically, by assuming *given* individuals, the microfoundations project in orthodox economics had typically to assume furthermore that each and every individual was *identical* in order to attempt to make the analysis tractable. The concept of an institution, properly handled, points not to a spurious supraindividual objectivity, nor to the uniformity of individual agents, but to the concept of socioeconomic order, arising not despite, but because of, the variety at the microlevel. Without such microvariety, there would be no evolutionary development of the processes of conformism and emulation that can sustain order.

Generally, institutions fill a key conceptual gap in social and economic theory. Institutions simultaneously constitute, and are constituted by, human action. Institutions are sustained by "subjective" ideas in the heads of agents, and are also "objective" structures faced by them. Choosing institutions as important units of analysis does not imply that the role of the individual is surrendered to the dominance of institutions. Both individuals and institutions are mutually constitutive of each other. Accordingly, explanations cannot be reduced to institutions alone, nor to individuals alone.[10]

Full, phylogenetic evolution in the socioeconomic sphere involves the evolution of individuals—their capacities, purposes, and preferences—as well as of institutions. Institutionalists, from Veblen to John Kenneth Galbraith, have always stressed the way in which institutions both mould and constrain human cognition and behavior (Hodgson 2000). However, this does not mean that there is no scope for human agency.

The institutionalist John Commons (1934 p. 69) noted that: "Sometimes an institution seems analogous to a building, a sort of framework of laws and regulations, within which individuals act like inmates. Sometimes it seems to mean the 'behavior' of the inmates themselves." This dilemma of viewpoint persists today. For example,

---

[10] Contrary to Lawson in this volume, I am opposed to any such reductionist explanations.

Douglass North's (1990 p. 3) definition of institutions as "rules of the game . . . or . . . humanly devised constraints" stresses the restraints of the metaphorical prison in which the "inmates" act. In contrast, Veblen's (1919 p. 239) definition of an institution as "settled habits of thought common to the generality of men" seems to start not from the objective constraints, but from "the inmates themselves." However, as Commons himself concluded, the thrust of the "old" institutionalist approach is to see behavioral habit and institutional structure as mutually entwined and mutually reinforcing: both aspects are relevant to the full picture. A dual stress on both agency and structure is required.

The proposed alternative is not a crude holism. Complete explanations of parts in terms of wholes are beset with problems of equivalent stature to those of the inverse procedure. Just as structures cannot be adequately explained in terms of individuals, individuals cannot adequately be explained in terms of structures. Fortunately, there are sophisticated alternative approaches in philosophy and social theory (Archer 1995; Bhaskar 1979; Giddens 1984; Kontopoulos 1993) that emphasise the structured interaction of parts with wholes, and eschew single-level explanations.

6 In Conclusion: Time Past and Time Future

It should now be evident how the three core attributes of an "evolutionary economics"—that of embracing novelty, of opposing reductionism, and of incorporating broad and phylogenetic conceptions of change—are linked. Underling the first two is the concept of emergence. This concept provides a means of (a) reconciling causal explanations with purposeful and *seemingly* "uncaused" behavior, and (b) establishing relatively autonomous levels of analysis above, and in addition to, the microeconomic level. The third idea of incorporating broad, phylogenetic conceptions of evolution in which there is both change in, and variety of, human purposes and preferences is now seen as necessary to the evolution of the economic institutions and order which are the real foundation of (b). Evolutionary economics may thus take advantage of an integrated package of concepts and approaches involving all three core attributes.

A number of developments in modern economics are complementary to this. Gary Becker (1962), in an early and uncharacteristic article, and more recently Dhananjay Gode and Shyam Sunder (1993) have constructed models where systemic constraints prevail over microvariations. Becker demonstrated that an "irrational" mode of behavior, in which agents are ruled by habit and inertia, is just as capable of predicting the standard downward-sloping demand curve and the profit-seeking activity of firms. Becker showed how the negatively inclined market demand curve can result from habitual behavior. Actors "can be said to behave not only 'as if' they were rational but also 'as if' they were irrational: the major piece of empirical evidence justifying the first statement can equally well justify the second" (Becker 1962 p. 4). Kenneth Arrow (1986) has also accepted the possibility of an alternative approach based on habit. Gode and Sunder go on to show that experiments with agents of "zero intelligence" produce predictions that differ little from those with human traders. Two conclusions follow. First, the "accuracy of the predictions" or other familiar criteria for theory selection do not give outright victory to rational choice models. Second, these models suggest that ordered and sometimes predictable behavior can result from *institutional constraints*, and may be largely independent of the "rationality," or otherwise, of the agents.

Other allied inspirations can be gained from complexity theory. The literature on complex systems and emergent properties lends support to the "old" institutionalist idea that the economy can and must be analysed at different levels (Lane 1993; Cohen and Stewart 1994). There is a valid and sustainable distinction between the "micro" and the "macro," without reducing the former to the latter, or vice-versa. The concept of an institution provides a key conceptual bridge between the two levels of analysis. The concept of an institution connects the microeconomic world of individual action, of habit and choice, with the macroeconomic sphere of seemingly detached and impersonal structures. While analyses at each level must remain consistent with each other, the macroeconomic level has distinctive and emergent properties of its own.

Such alternative approaches—of which only a most rudimentary and preliminary sketch has been provided here—do not always lend themselves to formal modeling. Complexity itself imposes limits on mathematical modeling and formal theory. The adoption of more modest formal aims and the shifting of the balance of economic theory away from mathematics may be a necessary outcome. Indeed, the problems faced by the microfoundations project in mainstream economics suggest that there are limits to the deductivist conception of theory that has dominated economics for the last 50 years. As Frank Hahn (1991 p. 50) has conjectured: "Not only will our successors have to be far less concerned with . . . grand unifying theory . . . [but also] less frequently for them the pleasures of theorems and proof. Instead the uncertain embrace of history and sociology and biology." Evolutionary economics can rise to this challenge. It is all the stronger in doing so if it has a clear understanding of its own philosophical underpinnings and of its own past.

References

Alchian AA (1950) Uncertainty, evolution and economic theory. J Polit Econ 58:211–222. Reprinted in Witt (1993)

Andersen ES (1994) Evolutionary economics: post-Schumpeterian contributions. Pinter, London

Archer MS (1995) Realist social theory: the morphogenetic approach. Cambridge University Press, Cambridge

Arrow KJ (1986) Rationality of self and others in an economic system. J Bus 59:S385–S399. Reprinted in Hogarth and Reder (1987), and in Eatwell J, Milgate M, Newman P (eds) (1987) The new Palgrave dictionary of economics, 4 vols, vol 2. Macmillan, London

Axelrod RM, Hamilton W (1981) The evolution of cooperation. Science 211:1390–1396

Becker GS (1962) Irrational behavior and economic theory. J Polit Econ 70:1–13. Reprinted in Caldwell BJ (ed) (1993) The philosophy and methodology of economics, vol 1. Edward Elgar, Aldershot

Bhaskar R (1979) The possibility of naturalism: a philosophic critique of the contemporary human sciences. Harvester, Brighton

Bonaccorsi A, Pammolli F, Tani S (1995) On R&D and the nature of the firm. University of Pisa, mimeograph

Boulding KE (1978) Ecodynamics: a new theory of societal evolution. Sage, Beverly Hills
Boulding KE (1981) Evolutionary economics. Sage, Beverly Hills
Boulding KE (1991) What is evolutionary economics? J Evol Econ 1:9–17
Chiaromonte F, Dosi G (1993) Heterogeneity, competition, and macroeconomic dynamics. Struct Change Econ Dyn 4:39–63
Clark JB (1885) The philosophy of wealth: economic principles newly formulated. Macmillan, London, New York
Cohen J, Stewart I (1994) The collapse of chaos: discovering simplicity in a complex world. Viking, London, New York
Commons JR (1934) Institutional economics—its place in political economy. Macmillan, New York. Reprinted 1990 with a new introduction by M. Rutherford. Transaction, New Brunswick
Degler CN (1991) In search of human nature: the decline and revival of Darwinism in American social thought. Oxford University Press, Oxford, New York
Dosi G, Freeman C, Nelson R, Silverberg G, Soete LLG (eds) (1988) Technical change and economic theory. Pinter, London
Elster J (1982) Marxism, functionalism and game theory. Theory Soc 11:453–482. Reprinted in Roemer JE (ed) (1986) Analytical Marxism. Cambridge University Press, Cambridge
Foss NJ (1994) Realism and evolutionary economics. J Soc Evol Syst 17:21–40
Foster J (1997) The analytical foundations of evolutionary economics: from biological analogy to economic self-organisation. Struct Change Econ Dyn 8:427–451
Freeman C (ed) (1990) The economics of innovation. Edward Elgar, Aldershot
Friedman M (1991) Old wine in new bottles. Econ J 101:33–40
Georgescu-Roegen N (1971) The entropy law and the economic process. Harvard University Press, Cambridge
Giddens A (1984) The constitution of society: outline of the theory of structuration. Polity Press, Cambridge
Gode DK, Sunder S (1993) Allocative efficiency of markets with zero-intelligence traders: market as a partial substitute for individual rationality. J Polit Econ 101:119–137
Hahn FH (1991) The next hundred years. Econ J 101:47–50
Hamilton WH (1932) Institution. In: Seligman ERA, Johnson A (eds) Encyclopaedia of the social sciences, vol 8. Macmillan, New York, pp 84–89. Reprinted in Hodgson GM (ed) (1993) The economics of institutions. Edward Elgar, Aldershot
Hanusch H (ed) (1988) Evolutionary economics: applications of Schumpeter's ideas. Cambridge University Press, Cambridge
Hayek FA (1982) Law, legislation and liberty, 3-vol combined edn. Routledge and Kegan Paul, London
Hayek FA (1988) The fatal conceit: the errors of socialism. The collected works of Friedrich August Hayek, vol I. Bartley WW III ed. Routledge, London
Heertje A (1994) Neo-Schumpeterians and economic theory. In: Magnusson L (ed) (1994) Evolutionary and neo-Schumpeterian approaches to economics. Kluwer, Boston, pp 265–276
Hobhouse LT (1913) Development and purpose: an essay towards a philosophy of evolution, 1st edn. Macmillan, London
Hodgson GM (1988) Economics and institutions: a manifesto for a modern institutional economics. Polity Press and University of Pennsylvania Press, Cambridge, Philadelphia
Hodgson GM (1993) Economics and evolution: bringing life back into economics. Polity Press and University of Michigan Press, Cambridge, Ann Arbor
Hodgson GM (1997) Economics and evolution and the evolution of economics. In: Reijnders J (ed) (1997) Economics and evolution. Edward Elgar, Cheltenham, pp 9–40
Hodgson GM (1998) On the evolution of Thorstein Veblen's evolutionary economics. Cambridge J Econ 22:415–431
Hodgson GM (1999) Evolution and institutions: on evolutionary economics and the evolution of economics. Edward Elgar, Cheltenham

Hodgson GM (2000) What is the essence of institutional economics? J Econ Issues 34:317–329
Hutter M (1994) Organism as a metaphor in German economic thought. In: Mirowski, 1994, pp 289–321
Kontopoulos KM (1993) The logics of social structure. Cambridge University Press, Cambridge
Lachmann LM (1969) Methodological individualism and the market economy. In: Streissler EW (ed) (1969) Roads to freedom: essays in honour of Friedrich A. von Hayek. Routledge and Kegan Paul, London, pp 89–103. Reprinted in Lachmann, 1977
Lachmann LM (1977) Capital, expectations and the market process, edited with an introduction by Grinder WE. Sheed Andrews and McMeel, Kansas City
Lane DA (1993) Artificial worlds and economics. J Evol Econ 3:89–107, 177–197
Lawson T (1997) Economics and reality. Routledge, London
Leontief WW (1982) Letter in Science 217:104, 107
Loasby BJ (1976) Choice, complexity and ignorance: an enquiry into economic theory and the practice of decision making. Cambridge University Press, Cambridge
Marshall A (1890) Principles of economics: an introductory volume, 1st edn. Macmillan, London
Mayr E (1985) How biology differs from the physical sciences. In: Depew DJ, Weber BH (eds) (1985) Evolution at a crossroads: the new biology and the new philosophy of science. MIT Press, Cambridge, pp 43–63
McDougall W (1929) Modern materialism and emergent evolution. Methuen, London
Menger C (1981) Principles of economics. Edited by J. Dingwall and translated by B.F. Hoselitz from the German edition of 1871. New York University Press, New York
Mirowski P (1989) More heat than light: economics as social physics, physics as nature's economics. Cambridge University Press, Cambridge
Mirowski P (ed) (1994) Natural images in economic thought: markets read in tooth and claw. Cambridge University Press, Cambridge, New York
Mitchell WC (1937) The backward art of spending money and other essays. McGraw-Hill, New York
Morgan CL (1896) Habit and instinct. Edward Arnold, London, New York
Morgan CL (1923) Emergent evolution, 1st edn. Williams and Norgate, London
Morgan CL (1933) The emergence of novelty. Williams and Norgate, London
Murphy JB (1994) The kinds of order in society. In: Mirowski, 1994, pp 536–582
Nelson RR (1995) Recent evolutionary theorizing about economic change. J Econ Lit 33:48–90
Nelson RR, Winter SG (1973) Towards an evolutionary theory of economic capabilities. Am Econ Rev (Papers and Proceedings) 63:440–449
Nelson RR, Winter SG (1974) Neoclassical vs. evolutionary theories of economic growth: critique and prospectus. Econ J 84:886–905. Reprinted in Freeman, 1990
Nelson RR, Winter SG (1982) An evolutionary theory of economic change. Harvard University Press, Cambridge
North DC (1990) Institutions, institutional change and economic performance. Cambridge University Press, Cambridge
Popper KR (1982) The open universe: an argument for indeterminism. From the postscript to the Logic of Scientific Discovery, Bartley WW III (ed) Hutchinson, London
Rizvi SAT (1994) The microfoundations project in general equilibrium theory. Cambridge J Econ 18:357–377
Schumpeter JA (1934) The theory of economic development: an inquiry into profits, capital, credit, interest, and the business cycle. Translated by R. Opie from the second German edition of 1926, first edition 1911. Harvard University Press, Cambridge. Reprinted 1989 with a new introduction by J. E. Elliott. Transaction, New Brunswick
Schumpeter JA (1942) Capitalism, socialism and democracy, 1st edn. George Allen and Unwin, London

Schumpeter JA (1954) History of economic analysis. Oxford University Press, New York
Shackle GLS (1955) Uncertainty in economics. Cambridge University Press, London
Shackle GLS (1989) What did the "general theory" do? In: Pheby J (ed) (1989) New directions in post-Keynesian economics. Edward Elgar, Aldershot, pp 48–58
Veblen TB (1898) Why is economics not an evolutionary science? Q J Econ 12:373–397. Reprinted in Veblen, 1919
Veblen TB (1899) The theory of the leisure class: an economic study in the evolution of institutions. Macmillan, New York
Veblen TB (1919) The place of science in modern civilisation and other essays. Huebsch, New York. Reprinted 1990 with a new introduction by WJ Samuels. Transaction, New Brunswick
Veblen TB (1934) Essays on our changing order. L Ardzrooni editor. Viking Press, New York
Vining R (1949) Methodological issues in quantitative economics. Rev Econ Stat 31(2):77–86
Williamson OE, Winter SG (eds) (1991) The nature of the firm: origins, evolution, and development. Oxford University Press, Oxford, New York
Winter SG Jr (1982) An essay on the theory of production. In: Hymans SH (ed) (1982) Economics and the world around it. University of Michigan Press, Ann Arbor, pp 55–91
Winter SG Jr (1988) On Coase, competence, and the corporation. J Law Econ Organ 4:163–180. Reprinted in Williamson and Winter, 1991
Winter SG Jr (1990) Survival, selection, and inheritance in evolutionary theories of organization. In: Singh JV (ed) (1990) Organizational evolution: new directions. Sage, London, pp 269–297
Witt U (1987) Individualistiche Grundlagen der evolutorischen Ökonomie. Mohr, Tübingen
Witt U (ed) (1992) Explaining process and change: approaches to evolutionary economics. University of Michigan Press, Ann Arbor
Witt U (ed) (1993) Evolutionary economics. Edward Elgar, Aldershot

Economics and Darwinism*

ULRICH WITT

Summary. The dramatic change that has occurred in the economy over the past few centuries is now accelerating. In view of this, an evolutionary approach to economic theorizing is now seriously being considered. If such an approach is adopted in economics, what role may be played by Darwinian thought on evolution as developed in evolutionary biology? This is the question to be discussed in this chapter. Evolutionary economics has indeed made recourse to Darwinian concepts, though mostly in a metaphorical way or by the construction of analogies to the Darwinian core notion of natural selection. Alternatively, a more direct approach may be suggested in which Darwinian thought is applied to the explanation of human economic behavior. As will be argued, the latter position may indeed be extended to explain significant historical features of long-term economic development in a broader Darwinian world view. With respect to the preference subjectivism that has prevailed in economics for a century, such a change in perspective may be a sign of a turnaround.

Key words. Darwinian world view, Evolutionary economics, Natural selection analogy, Adaptation, Preferences

1 Introduction

Current theorizing in economics is mainly directed toward two goals. First, there is the functionalist endeavor to explain, or better rationalize, why certain kinds of behavior and institutions can be observed in the economy. Second, efforts are made to attribute, in a hypotheticonormative manner, certain features such as efficiency, equity, or optimality to the outcome of the economic process. Both strands of thought relate to a definition of economics as a discipline which deals with the problem of scarcity. As a result of "nature's parsimony," as Ricardo once put it, humans (and prob-

---

Max Planck Institute for Research into Economic Systems, Evolutionary Economics Unit, Kahlaische Str. 10, D-07745 Jena, Germany

* Originally published as "Evolutionary Economics and Evolutionary Biology" in *Sociobiology and Bioeconomics*, edited by P. Koslowski (Springer, Berlin and Heidelberg, 1999), 279–298. With the permission of the publisher.

ably not only humans, see Ghiselin 1978) have to use scarce resources to achieve given ends. An alternative to this latently normative orientation is to interpret economics as a theory of human social behavior in the context of what are usually considered economic activities, namely production, accumulation, distribution, exchange, and consumption. From this perspective, explaining why and how the historically observable forms of production, accumulation, distribution, exchange, and consumption change so significantly, and why the changes have accelerated so much in the last few centuries, is a natural goal for economic theorizing.

It is precisely this reinterpretation of the economic sphere as undergoing continuous change which has recently induced increasing efforts to develop an evolutionary approach to economics. Some economists at the turn of the century, who had already been influenced by the "Darwinian revolution" in natural history and the sciences in the 19th century, had already expressed an interest in such an approach.[1] However, their pleas did not have much influence on actual economic theorizing. Darwin's—and even more Spencer's—thought seemed to have been inspired by the idea of extending the upper classes' liberal social philosophy of Victorian Britain to nature (Young 1988; Bowler 1989, 1995; Desmond and Moore 1991; Richards 1992). The inspiration guiding economic theorizing at the end of the 19th century was quite different. The proponents of the "neoclassical revolution" in economic theory (Walras 1954; Edgeworth 1881; Pareto 1896) tried to establish economic theory in terms of concepts borrowed from classical mechanics (see Mirowski 1988). In this, they were turning their backs on both the pre-Victorian (Smith 1759, 1776) and the Victorian (Mill 1848) versions of social philosophy. They had a "mechanics of utility and self-interest" in mind in which the state of rest, and not processes of change, figured most prominently. The focus was on a (market) equilibrium in which not only all individual plans fit together, but in which all the agents' utility reaches an extreme value, a maximum in this case, subject to the mutually imposed constraints. (This is the dual formulation of the concept of equilibrium in classical mechanics where the state of rest is associated with minimum free energy.) The basic concepts and the way in which they were presented came from Newtonian physics—the approach which the "Darwinian revolution" had just begun to challenge (Mayr 1991 Chap. V).

In view of these markedly different intellectual developments, and the fact that an evolutionary approach is now being used more seriously in economics, it may be asked how relevant the "Darwinian revolution" is for economics. In economics, attempts actually to make recourse to Darwinian concepts have so far been undertaken in two different ways. One is the direct application of the theory of natural selection to human economic behavior. The other is a metaphorical use of Darwinian concepts in the construction of analogies. As is to be expected, the different approaches imply entirely different answers to the question just posed. This chapter briefly outlines these approaches and discusses possible parallels with evolutionary biology. Section 2 focuses on attempts to apply the theory of natural selection directly to the economic domain. Section 3 turns to contributions in economics in which the theory of natural selection is used as a metaphor. In Sect. 4, the theoretical background of a more indirect interpretation is sketched. Section 5 highlights some of the significant historical

[1] See the often-cited introduction to Marshall (1938), Veblen (1898), Schumpeter (1934), a good survey in Hodgson (1993), and a recent restatement of such interest in Khalil (1992).

features of long-term economic development, and argues that the explanatory sketch given previously is helpful in making sense of the observed features of economic evolution in terms of a broader Darwinian world view. Section 6 offers some conclusions.

2 Does Natural Selection Operate on Economic Behavior?

Attempts to extend Darwinism directly to human economic behavior have been made in contributions covering topics such as game theory (Hansson and Stuart 1990; Güth and Yaari 1992) and social philosophy (Hayek 1967, 1979, 1988). The game-theoretic contributions are motivated by the fact that humans are a result of genetic evolution and that human behavior, in particular all kinds of behavior conducive to, or instrumental in, enhancing reproductive success, has been subject to natural selection. Following the lead of sociobiology in explaining the emergence of "reciprocal altruism" (Trivers 1971), evolutionary game theory has been applied to social dilemmas, particularly the prisoners' dilemma, arising in economic interactions. Conditions are derived under which, in a suitably defined human population, agents who share "altruistic" genes have a selective advantage. They behave cooperatively even when, with self-interested preferences, rationality forbids such behavior. The contributions provide a logically sound explanation for the compatibility of altruism and rational decision making within a certain population where otherwise these are irreconcilable. It is the irreconcilability that gives rise to the dilemma. Taken by itself, this application of evolutionary game theory to human behavior can be interpreted as referring to early stages of human phylogeny during which selection pressure was intense. Preferences for behaving altruistically towards certain persons may then have been genetically fixed, and thus still be present today. This argument has something to it, and will be discussed further in Sect. 4 in a more general form.

The argument does not imply that the process described by evolutionary game theory is still operating on human behavior today. In fact, such an interpretation would be much stronger and difficult to defend. A vast array of inherited traits determines the potential behavior of each individual in a population of genetically reproducing organisms of a certain species. Genetic variety in the gene pool of the population translates into a distribution of traits, and thus into potential behavior.[2] The case in which differences in the inherited traits of the organisms imply differential reproductive success and induce a change in the distribution of traits in the gene pool of the next generation is a situation of selection pressure. In the original, Malthusian, spirit of Darwin's theory, selection pressure may be considered to reflect a situation of competition between organisms for reproductive success or for items that are

[2] On the ontogenetic level, environmental influences (e.g., nutrition conditions) which affect the growth of the phenotypic organism may differ in their impact on actual behavior and thus affect the extent to which this behavior lives up to its genetically determined potential. However, the ontogenetic differences are supposed to have a minor and nonsystematic influence on reproductive performance, i.e., the number of living offspring who carry the gene pool of the next generation, compared with the influence of genetic differences.

conducive to that success.[3] In the absence of selection pressure—in a situation of affluence—behavior which contributes to genetic fitness is no longer selected for. Behavioral variety may then increase and include variants which have no adaptive value. Thus, in a state of affluence, the Darwinian approach to explaining behavior by reference to its contribution to inclusive fitness is much less significant than under conditions of selection pressure. Given the abundance which modern economies have been able to create, it can scarcely be doubted that human behavior can vary significantly without having a differential impact on reproductive success. In particular, economically more or less efficient forms of behavior do not seem to correlate with the generation of more or fewer offspring, not even when a trade-off between quality and quantity in raising offspring is accounted for.[4]

In a modified form, a direct application of Darwinian thought also underlies the theory of societal evolution outlined by Hayek.[5] He assumes three levels of evolution. The first is that of genetic evolution in which primitive forms of social behavior, preferences, and attitudes which bring about order in social interactions have been fixed genetically during man's phylogeny. Second, there is the evolution of the products of human intelligence and knowledge. Freed from the finite existence of each individual brain by efficient forms of coding, storing, and transmitting information, human knowledge has expanded enormously. Today, the extent to which it makes the mastery of nature possible is impressive. Third, and this is what Hayek considers the core of his approach, there is the level of cultural evolution operating "between instinct and reason" (Hayek 1988 Chap. 1). Culture, in Hayek's interpretation, is neither genetically conditioned nor rationally designed. It is a tradition of learnt "rules of conduct" (Hayek 1967) whose role, he claims, is often not understood even by those who follow them. They are passed on through cultural transmission—a "blind" or spontaneous process in the sense that it is not consciously planned or controlled.

Hayek argues that the question of which rules come into being, and how this happens, is a matter of historical accident. In contrast, systematic influences that result from a selection process decide which of the rules survive. He believes that the process shapes cultural evolution and operates on groups, where the groups are defined as sharing the same rules of conduct. Those groups which succeed in developing and passing on rules better suited to governing their social interactions are able to grow and to feed a larger number of people. Their relative superiority may enable them to conquer and/or absorb less well-equipped, competing groups and thus, unintentionally, propagate the superior sets of rules. A growing population requires increasing specialization and division of labor. This, in turn, means that the underlying rules of conduct and the order they induce are extending. These rules then become more differentiated and abstract (and, for the single mind, more difficult to understand). An

---

[3] Such as, for example, the size and quality of a habitat. It is worth noting that competition for scarce resources may exert its influence through an ecological effect. Competition for feeding opportunities at the level of the predator may induce selection pressure at the level of the prey if the adaptive features of the prey population vary, as in the case of the evolution of industrial melanism.

[4] For further discussions, see Witt (1985) and Hermann-Pillath (1991).

[5] For statements of various aspects of that theory see the epilogue in Hayek (1979) in which he explicitly draws on older notions of group selection in sociobiology; see also Hayek (1988).

"extended order" is seen as spontaneously emerging in this way over centuries. It is what has enabled modern societies to achieve a historically unique level of civilization and productivity. This order embodies an impersonal intelligence accumulated during the selection processes in the form of surviving impersonal rules of conduct. Its most important achievement, Hayek (1988 Chap. 3) submits, is trade and the emergence of a system of markets.

Thus, Hayek's theory is based on the twin ideas of spontaneous order as a system of impersonal rules of conduct and evolution as a result of group selection. Taken together, they amount to claiming that natural selection not only "chooses" between competing species, it also "chooses" between competing groups of humans in which acquired cultural norms and rules are fixed. The problem with Hayek's grand view is that it offers no explanation of the crucial processes of social learning and the transmission of the rules of conduct. How do these processes operate? Is their continuity sufficient to ensure that group selection can indeed exert a differential effect on the replication of the rules? Since much of the process is tacit, is it possible that "observational" or "vicarious" learning (Bandura 1986), i.e., learning by imitation, plays a crucial role? How does this fit the standard interpretation of group selection? These questions do not deny the possibility that social or collective learning, which takes place within stable groups of agents, is an essential element of economic evolution and may yield a basis for deriving appropriate transition laws (for a corresponding modeling effort see, e.g., Witt 1996). The point is, rather, that these processes of transmitting knowledge within a human population are too little understood to justify the far-reaching conclusions Hayek wants to draw.[6]

3 Natural Selection as a Metaphor

Attempts to construct analogies between genetic and economic evolution (e.g., in Boulding 1981; Hirshleifer 1982; Matthews 1984; Allen 1988; Metcalfe 1989; Faber and Proops 1990; Saviotti and Metcalfe 1991; Hodgson 1995; Anderson 1994; Gowdy 1994 Chap. 5) seem to be more popular than attempts to apply the theory of natural selection directly, or so it seems from their frequency. Yet there are serious objections to this kind of transfer too. To demonstrate this, imagine, for the moment, that the metaphor is used in the economic context. What, then, would the objects of economic selection be, or what would have to be thought of as the evolving entity in the domain of economics? Quite diverse answers to these questions have been suggested in the literature. Some studies, for instance, focus on human artefacts such as mechanical devices for power transmission (Usher 1954), or products which serve certain particular needs (Saviotti 1988). The history of such artefacts is often well documented. The

[6] The coevolutionary approach suggested, among others, by Boyd and Richerson (1985) covers several ideas of Hayek's theory in detail, and is quite explicit in deriving corresponding transition laws. However, where it is not based on analogy, its crucial assumption is that social learning processes and genetic evolution interact or interfere with one another. Implicitly, this view presupposes that the theory of natural selection can be applied, albeit in a biased form, to present-day human behavior, an interpretation that has been rejected here for the reasons given above.

Comptoise clock, which was produced from 1680 to 1914 in a small part of eastern France, is a telling example (see Schmitt 1983).

The basic mechanical layout of that pendulum clock has changed in only minor ways. The transition from manufacturing by hand (an estimated 20000 pieces in roughly the first hundred years) to mechanized mass production (an estimated 2 million pieces in the following hundred years) has been accompanied mainly by changes in the outer appearance of the clock and the materials used. There have been several generations of faces, pendulums, and devices for striking the hour. The evolution of this artefact can thus be seen in the changes in its production technology, its mechanical layout, and its appearance. Its "population dynamics" were significant before it eventually became extinct. Nonetheless, it is hard to provide a coherent explanation of the development in terms of a selection model. Many of the changes made may be due to limitations on the materials available or to manufacturers' costs. Other changes may have been introduced by the manufacturers in anticipation of changes in fashion, i.e., biased adoption decisions by the consumers. No general causal structure seems to underlie these changes, and nor is there any hint of a continuously operating selection mechanism. The development of other artefacts has had different features, so that the chances of finding a coherent explanation on the basis of the selection metaphor appear negligible.[7]

Another answer to the question of what might be the object of economic selection, or the evolving entity in the domain of economics, has been given in the literature on firms' organizations and their growth, persistence, decline, and eventual exit from the market. These contributions grew out of the debate on "economic natural selection," which has been instrumental in the development of evolutionary economics.[8] The question raised was whether, and to what extent, diversity in the firms' goals and performances would be eliminated by competition. Would all those forms which are not sufficiently profitable be driven out of the market? The idea was explicitly derived by analogy with Darwinian thought. However, for logical reasons, the analogy is difficult to defend. First, no convincing analogue for the two separate, systematically interacting, levels of the genotype and the phenotype can be found in the economic domain, a point already made by Winter (1964). Second, there is no entity comparable to a gene pool in terms of continuity or, more specifically, no equivalent to the continuity of the germ plasma. As a consequence, the notions of selection and replication refer to much less constrained and structured processes than in the Darwinian example. In evolutionary economics they therefore lack much of the significance they have in evolutionary biology.

Despite these reservations, the weakened form of the analogy suggested by Nelson and Winter (1982) is, up to the present, probably the most influential contribution to evolutionary economics. They argue that firms are organizations which have to base

[7] Several aspects of the changes which occurred in this and many other artefacts may, however, be captured quite well by a diffusion approach not based on any notion of selection: see Rogers (1983), Metcalfe (1988), and Nakicenovic and Grübler (1991). As explained in more detail elsewhere (Witt 1992), the frequency–dependency effect is a dominant feature of diffusion processes and represents an important element in formulating respective transition laws.

[8] Alchian (1950), Penrose (1952), Friedman (1953); among the few empirical studies are Wedervang (1965) and, more recently, Audretsch (1995).

their internal interactions on behavioral routines, rules of thumb, and regular interaction patterns. The firms' behavior in production planning, calculation, price setting, and even the allocation of R&D funds is viewed as rule-bounded. The corresponding "routines" are interpreted as the "genotypes," while the firm's specific decisions thus derived are interpreted as "phenotypes." The latter may be more or less favorable for the firm's overall performance measured in terms of profitable growth. Assuming that routines which successfully contribute to growth will not be changed, the actual expansion can be understood as an increase in relative frequency of those "routine-genes." Routines which cause a deterioration in the firm's performance, on the other hand, are unlikely to replicate. The development of behavioral routines is thus considered to be a significant instance of economic evolution and an object of selection processes. Yet there is little to support the assumption that this development is governed by a continuously operating selection mechanism (perhaps even one following transition laws given by a time-dependent change of respective frequency distributions). On the contrary, it is most likely that firms facing deteriorations are induced to identify the deficient routines and to replace or improve them in a kind of intentionally produced mutation of their "genes."

Nelson and Winter (1982) argue that such improvement or replacement activities are themselves subject to (higher) routines, and that the differential growth argument therefore applies once again. It may indeed be that firms activate higher problem-solving routines on a regular basis when a crisis occurs or is expected to occur. Yet the outcome of these problem-solving routines is often more likely to depend on whose problem-solving capacity is involved in those routines than on the routines themselves. In this case, the differential success of the problem-solving process triggered cannot be derived from differences in those routines. Differential success is basically unpredictable (much like mutation in the genetic sphere). When different people are involved in the same routines at different times, development may be different. The continuity of what has been suggested as the analogue of the selection mechanism is lost. The argument derived from the selection metaphor rests on shaky foundations.[9]

In a more general perspective, there is yet another reason for the limited usefulness of the selection metaphor in the economic domain. In the genetic context, selective forces operate on a given population and change the relative frequencies of the genes in the pool. The individual members of the population have little, if any, room for escaping from these pressures or for adjusting to them. In this sense, selection forces may be labeled "external." In a similar way, the anonymous competitive forces of the markets can be seen as external to the individual. The market participants unintentionally impose constraints on each other in their multilateral interactions, and these constraints may even force some of them out of business. Thus the population of

[9] The conclusion which Audretsch (1995 p. 186) draws from his empirical findings is worth noting in this context: "... the main message ... is that, with respect to the dynamic patterns of firms over time, there is, in fact, no tendency that can be generalized. Rather, the dynamic nature in which firms and industries tend to evolve over time varies substantially from industry to industry. And there is at least some evidence that it is differences in the knowledge condition and technology underlying the specific industry, that is the nature of innovative activity, that account for variations in industry evolution across markets."

agents on the supply and/or demand sides of a market may systematically change over time. Yet, unlike in the genetic case, the agents are not helpless when exposed to these changes. Indeed, an important part of the systematic changes observable in markets, both in the composition of the participants and in their activities, may be attributed to the individuals' attempts to anticipate the effects of the market forces and to take account of them. If successful, the effects of "external" selection are replaced by what would have to be labeled "internal" selection—an attempt to change activities deliberately and so to avoid unfavorable selection consequences.

Since this dimension of economic behavioral adaptation has no equivalent in the genetic domain, the selection metaphor may be a misleading heuristic for understanding the crucially important feature of economic evolution—the role played by learning and cognition. In fact, the individuals' capacity to anticipate the consequences of their behavior, and to take account of them deliberately by searching for better solutions is more akin to Lamarckian than to Darwinian evolution. This is the reason why economic activities are much more volatile. Nonetheless, adaptive features are present here as well, and they may be conjectured to follow their own regularities. Because of the limited human information-processing capacity, these regularities are likely to reflect mental, rather than genetic, selection processes—people are forced to be selective in what they sense, learn, and perceive. Since a coherent theory of these mental processes is still lacking, the present understanding of "internal" selection processes and problem solving leaves many questions open. In any case, it may be inferred from the insights already available that the anticipation of external selection forces by internal selection, i.e., cognitively guided behavioral adaptations, does not necessarily lead to the same outcome as external selection alone. Cognitive anticipation usually allows for a multiplicity of solutions to a problem. This is a forteriori true where the invention of new, as yet unknown, solutions is considered.

4 Economic Preferences and the Genetic Inheritance[10]

Even though there does not seem to be any evidence that natural selection forces govern the rapid adaptations of economic behavior observable in modern societies, they may have an indirect effect. During human phylogeny, natural selection did shape human genetic endowments which, because those selection forces are no longer a source of systematic change, are still present and thus may have an impact on economic behavior. Among these endowments the basic, apparently genetically coded, conditioning or learning mechanisms and their "primary reinforcers" (Skinner 1966) seem to be particularly relevant. Before explaining their influence on economic activities, it is necessary to elaborate briefly on the notion of reinforcers developed in psychology, and on the notion of preferences considered in economics, as a determinant of human decision making.

In the economists' view, individuals are supposed to order, according to their subjective preferences, alternatives which are available for choice. When a choice has to be made, people decide in favor of the most preferred of their alternatives. Under certain conditions, preferences can be represented by a utility function in which the

[10] For a more detailed discussion of the material of this section see Witt (1991).

independent variables are quantities of the objects of choice. Depending on the particular choices available, any objects can, in principle, enter the utility function. Since the individuals' subjective preference orderings are not directly observable, laboratory revelation procedures must be designed in which the ordering of a set of objects (as given by the experimentor) can be inferred indirectly from the observed choices. Although an individual's preferences may be viewed as reflecting the desires of that individual, economic theory usually refrains from attempting to specify, in a more general way, the objects from which utility is derived. With reference to the subjective nature of their evaluations, the question of whether individuals have certain preferences for objects in common, which would be expected when preferences have a genetic basis, is left open.

It is interesting now that, in developing a theory of revealed preferences, economists have tried to design basically the same experiments which behavioral psychologists have set up for a different purpose, namely when an attempt is made to determine reinforcers empirically, that is, those items which have the power to reinforce the operant behavior of an organism (Lea 1983). If a preference revelation experiment were run with mammals, some basic items could be expected to be found as objects of the animals' preferences—those which have been identified in laboratory experiments with mammals (Millenson 1967) as primary reinforcers: air, water, sleep, warmth, nutrition, sexual activity, maternal care, love and affection, physical activity, and novelty. Because these elements are held in common, and because of their obvious adaptive value, they may be considered part of the mammals' genetic program. To provide a hypothesis about the common elements of individual preferences, something which is missing in economic theory, these items may as well be considered part of man's genetic endowment.

However, besides these innate, primary reinforcers, behavioral psychology suggests the existence of so-called "secondary" or conditioned reinforcers. These are acquired in a conditioning or learning process which builds on the innate learning capacity shared by all vertebrates, including man (Skinner 1966; Pulliam and Dunford 1980 pp. 11–44). According to this theory, learning secondary reinforcers takes place by simple association with primary reinforcing events. If an originally neutral stimulus is systematically paired with a primary reinforcing stimulus, an association is established in such a way that the learned stimulus can eventually act as a substitute, at least for a limited time. A stimulus associated with conditioned reinforcers can also acquire reinforcing power in this way. It can easily be imagined, given man's large associative capacity, how an extremely developed hierarchy of learned reinforcers emerges from the very simple innate learning mechanism.

In the light of this approach, some instructive conclusions can be drawn with regard to human preferences. Their variety can obviously be explained as the result of a permanent learning process, a slow process of preference formation and change. In an individual lifetime history of conditioning, a more or less long chain of learned associations leads from the few innate preferences to those actually revealed in current economic choices. These choices do not necessarily increase genetic fitness—people may, for instance, prefer to enjoy increased consumption rather than to raise children. Yet their decisions are related, in a complex, idiosyncratic fashion, to basic preferences that once had an adaptive value. In addition to the individual history of interactions of primary and secondary reinforcers, the unique historical circumstances that make certain choices available to certain individuals at a certain time also, of course, contribute to the enormous variance in actually observable choices in (and outside) the markets. The question that remains is whether, in spite of all the variance resulting from the individual influences, it can be argued that a general tendency can arise in the grand total of human choices in a way that is significant for characterizing economic evolution, and, if so, whether such a tendency has something to do with the basic innate preferences of man.

5 Economic Evolution—Between Anthropocentrism and "Nature's View"

It has been argued in the previous sections that the Darwinian theory of natural selection is of little heuristic use for conceptualizing evolution in the economic domain. There may, however, be an indirect influence through some basic, innate, human preferences which play a role in organizing and directing the individuals' learning or preference formation processes. Before this question can be decided, it is useful to outline briefly some of the significant features of economic evolution which have been observed, particularly over the past centuries. An almost trivial fact to be mentioned first is the undiminished exponential growth of the human population. An important prerequisite for the economic subsistence needed to feed a growing population has always been the provision of labor for the production of the means which, in turn, are capable of satisfying the basic needs. However, it is only in the last four centuries that growing technological knowledge has made two things feasible at the same time—a large-scale substitution of human physical work through nonhuman energy sources, and a real boost in the production volume. Wind and water power, wood, coal, oil, natural gas, and uranium were tapped one after another.

Coincidentally, the price of energy per unit, measured, e.g., in mega-Joules, fell strongly relative to the wage rate. This means that, on average, one hour of human work was able to buy an increasing amount of energy. Furthermore, the mechanization of the transformation of materials at large increased output produced per hour worked, i.e., the general productivity of labor. However, since energy intensity grew faster than labor productivity, the energy input per unit of output, the "energy coefficient," increased. This is particularly obvious in the agricultural sector, where a direct comparison between "primitive" methods of cultivation based exclusively on human energy inputs and modern methods based on nonhuman energy inputs is possible (see Pimentel and Pimentel 1996 Tables 10.3, 10.4, and 10.7). Needless to say, the productivity increases in agriculture thus achieved were instrumental in feeding a growing number of people. In addition, the productivity increases in the agricultural sector improved the nutrition of the existing population and resulted in an increasing life expectancy, as recent research on the period since 1800 has shown (see Fogel 1986).

The strongly improved "terms of trade" between human labor and nature's products (nonhuman energy flows and, with the necessary accumulation of machinery, materials) explain a fact that may seem puzzling, at least to noneconomists. In their historical development, modern economies could afford to shift the bulk of their expenditure (or, to look at it the other way round, the bulk of their income creation) first from agriculture to the industrial sector and then on to the service industries. Nowadays, agriculture and the industrial sector together generate less than half of domestic income in all OECD countries. Because of the comparatively high input share of human work hours needed in producing services, these services have become relatively more expensive compared with agricultural and manufactured goods. At the same time, the demand for services does not seem to have been curbed by their increasing relative price in favor of an expansion of demand for agricultural and manufactured goods, perhaps because the demand for the latter goods is more quickly satiated than demand for services.

Indeed, another striking feature of economic evolution over the past few centuries turns up on the demand side of the markets, where mankind's consumption has grown continually in both absolute and per capita terms. In both the developing and the developed countries, this is accompanied by increasing differentiation of consumer goods and the generation of entirely new groups of products. Economic growth is no longer only a matter of capital accumulation, technical progress, and productivity increases. It is also a story of what can be sold in the consumer markets. Qualitative change in the array of final consumption goods, increasing mobility, internationalization of trade in formerly local specialties, and the soaring diversification of the service sector are only the most salient aspects of a deep-seated process of change. This change is not without regularities. Economists have long known that the demand for certain product groups, such as groceries, does not expand at the same rate as does income available for consumption—a sign of satiation despite all the differentiation efforts in those industries. For other product groups, such as housing space and, in particular, those products which directly or indirectly provide comfort, safety, improvements in health, mobility, and entertainment, the demand effects are exactly the opposite, i.e., they indicate that satiation is a long way from being reached. The latter products are, of course, precisely those provided by the economy's service sector.

Are all these empirical findings just accidental, or do they fit into some general causal patterns? Economists are rarely inclined to accept this as a question they feel compelled to find an answer for. If long-run economic development is considered at all from a theoretical point of view, it is usually treated as a process of economic growth that reflects an increase in the (subjective) utility which most of the agents involved enjoy (Romer 1986)—the anthropocentric view of a basically hedonistic economic discipline. From an evolutionary perspective, things may be interpreted differently. Such a perspective considers human kind as just another species, albeit one that has been able to enjoy a prolonged period of affluence because its enhanced problem-solving capacity just kept pace with the ecological problems created. As was explained in the previous section, human decision making can be seen as being based on a complex structure of preferences that have emerged over the individuals' lifetimes from a few innate, and therefore basically interindividually shared, dispositions. Those complex structures vary greatly, and reflect idiosyncratic learning histories. It may therefore be very difficult to trace back the choices made by an economic agent at a given time to the few innate preferences that the agent shares with other agents. Nonetheless, the only consistent and enduring effect that could, on average, be expected to show up, despite all the individual variance in the decision making of human kind over an extended period of time, may be precisely an expression of those basic preferences that were fixed in earlier times of human phylogeny.

Some of the features of the long-term economic evolution sketched above do seem to fit this conjecture well. Humans behave much like any other species that dominates its environment. They increase their population size and pay little attention to the ecological dangers a rapidly growing population generates, and they make efforts to enhance individual health and life expectancy. They subdue all rival species or drive them to extinction. Humans use their growing knowledge to find ways of avoiding the burden of physical work in making a living. As far as they can, they expand their consumption of primary reinforcing items—food, drinks, entertainment, mobility, living space, and so on—to a level of satiation. Before satiation is reached, suppliers of those items search for and create more sophisticated ways of consuming those items. The satiation features of some items, such as food, set more narrow limits on the expansion of consumption than others, such as the diverse services mentioned above. Since many of these services, in particular those relating to mobility, rely more heavily on energy-flow inputs than on material inputs, energy consumption is unlikely to reach satiation in the future. If the working hypothesis about the role of innate primary reinforcers or preferences suggested above holds, it may be argued that a Darwinian perspective on economic evolution is fruitful, and that it leads to important material conjectures about economic evolution—provided, however, that perspective does not build on misleading metaphors or misconceived applications of the theory of natural selection to the domain of economics.

6 Conclusions

Human economic activity has changed significantly over time and, apparently, is continuing to do so in an accelerating fashion. In view of this, the static (and latently normative) interpretation of the economy in modern economic theory is quite inapt. An "evolutionist revolution" does seem warranted. However, unlike the amazingly simple and yet powerful principles of genetic variation and natural selection which, as Darwin discovered, underlie the evolution of species, the regularities which economic evolution may follow seem to operate in a way that is far more complex. The idea of applying the theory of natural selection directly to explain the rapid adaptations in modern economic behavior cannot be upheld. There is simply no evidence for a systematic, differential, reproductive effect on certain kinds of economic activities. Metaphorical use of core notions of the Darwinian theory—selection, variation, and replication—in an economic context is no less problematic. Economists sometimes indulge themselves in speaking of selection forces metaphorically to highlight the outcome of competitive processes, particularly market processes. Agents in the markets compete with each other, and the relatively inefficient, incompetent, or simply hypertrophic, competitors lose and eventually go out of business (an economic, not a physical, extinction). However, the differences between the actual economic and biological processes referred to by the term selection, and the results produced by these processes, are so different that the metaphor is of limited value for economic theory. The simple reason is that, in the domain of economics, there is no structure comparable in its continuity to the genetic mechanisms that have led to the emergence of species in nature. An inspection of long-term economic development can, however, help identify certain regularities which can be explained by evolutionary concepts as long as these are defined in a broader sense. A theory of preference formation has been suggested that points to some genetically coded human dispositions which may have a weak but persistent effect on economic choices, and which would explain the observed regularities. Economics may thus be fruitfully approached on the basis of a Darwinian world view in which natural selection has been, but no longer is, the immediate shaping agent.

References

Alchian AA (1950) Uncertainty, evolution, and economic theory. J Polit Econ 58:211–221

Allen P (1988) Evolution, innovation, and economics. In: Dosi G, Freeman C, Nelson R et al. (eds) Technical change and economic theory. Pinter, London, pp 95–119

Anderson ES (1994) Evolutionary economics: post-Schumpeterian contributions. Pinter, London

Audretsch DB (1995) Innovation and industry evolution. MIT Press, Cambridge

Bandura A (1986) Social foundations of thought and action—a social cognitive theory. Prentice-Hall, Englewood Cliffs

Boulding KE (1981) Evolutionary economics. Sage, Beverly Hills

Bowler PJ (1989) Evolution—the history of an idea, rev edn. University of California Press, Berkeley

Bowler PJ (1995) Herbert Spencers Idee der Evolution und ihre Rezeption. In: Engels EM (ed) Die Rezeption von Evolutionstheorien im 19. Jahrhundert. Suhrkamp, Frankfurt

Boyd R, Richerson PJ (1985) Culture and the evolutionary process. University of Chicago Press, Chicago

Desmond A, Moore JR (1991) Darwin. Michael Joseph, London

Edgeworth FY (1881) Mathematical psychics. Paul, London

Faber M, Proops JLR (1990) Evolution, time, production and the environment. Springer, Berlin

Fogel RW (1986) Nutrition and the decline of mortalities since 1700: some preliminary findings. In: Engerman SL, Gallman RE (eds) Long-term factors in American economic growth. Chicago University Press, Chicago, pp 439–555

Friedman M (1953) The methodology of positive economics. In: Friedman M. Essays in positive economics. University of Chicago Press, Chicago, pp 3–43

Ghiselin MT (1978) The economy of the body. Am Econ Rev 68, Papers and Proceedings, pp 233–237

Gowdy J (1994) Coevolutionary economics: the economy, society and the environment. Kluwer, Boston

Güth W, Yaari M (1992) Explaining reciprocal behavior in simple strategic games: an evolutionary approach. In: Witt U (ed) Explaining process and change—approaches to evolutionary economics. University of Michigan Press, Ann Arbor, pp 23–34

Hansson I, Stuart C (1990) Malthusian selection of preferences. Am Econ Rev 80:529–544

Hayek FA (1967) Notes on the evolution of systems of rules of conduct. In: Hayek FA (ed) Studies in philosophy and economics. Routledge and Kegan Paul, London, pp 66–81

Hayek FA (1979) The three sources of human values. Epilogue to: law, legislation, and liberty. Routledge, London

Hayek FA (1988) The fatal conceit. Routledge, London

Hermann-Pillath C (1991) A Darwinian framework for the economic analysis of institutional change in history. J Soc Biol Struct 14:127–148

Hirshleifer J (1982) Evolutionary models in economics and law. Res Law Econ 4:1–60

Hodgson GM (1993) Economics and evolution: bringing life back into economics. Polity Press, Cambridge
Hodgson GM (1995) The evolution of evolutionary economics. Scott J Polit Econ 82:469–488
Khalil EL (1992) Economics and biology: eight areas of research. Methodus 4:29–45
Lea SEG (1983) The analysis of need. In: Mellgren RL (ed) Animal cognition and behavior. North-Holland, Amsterdam, pp 31–63
Marshall A (1938/1898) Principles of economics, 8th edn. Macmillan, London
Matthews RCO (1984) Darwinism and economic change. In: Collard DA, Helm DR, Scott MFG et al. (eds) Economic theory and Hicksian themes. Clarendon Press, Oxford, pp 91–117
Mayr E (1991) One long argument. Harvard University Press, Cambridge
Metcalfe S (1988) The diffusion of innovations: an interpretative survey. In: Dosi G, Freeman C, Nelson R et al. (eds) Technical change and economic theory. Pinter, London, pp 560–589
Metcalfe S (1989) Evolution and economic change. In: Silberston (ed) Technology and economic progress. Macmillan, London, pp 54–85
Mill JS (1848) Principles of political economy with some of their applications to social philosophy. Parker, London
Millenson JR (1967) Principles of behavioral analysis. Prentice Hall, New York
Mirowski P (1988) Against mechanism—protecting economics from science. Rowman & Littlefield, Totowa
Nakicenovic N, Grübler A (1991) Diffusion of technologies and social behavior. Springer, Berlin
Nelson RR, Winter SG (1982) An evolutionary theory of economic change. Harvard University Press, Cambridge
Pareto V (1896) Cours d'èconomie politique. Rouge, Lausanne
Penrose ET (1952) Biological analogies in the theory of the firm. Am Econ Rev 42:804–819
Pimentel D, Pimentel M (1996) Food, energy, and society, rev edn. University of Colorado Press, Niwot
Pulliam HR, Dunford C (1980) Programmed to learn: an essay on the evolution of culture. Columbia University Press, New York
Richards RJ (1992) The meaning of evolution: the morphological construction and ideological reconstruction of Darwin's theory. Chicago University Press, Chicago
Rogers EM (1983) Diffusion of innovations, 3rd edn. Free Press, New York
Romer PM (1986) Increasing returns and long-run growth. J Polit Econ 94:1002–1037
Saviotti PP (1988) Information, variety and technoeconomic development. Res Pol 17:89–103
Saviotti PP, Metcalfe JS (1991) Present developments and trends in evolutionary economics. In: Saviotti PP, Metcalfe JS (eds) Evolutionary theories of economic and technological change. Harwood Academic Publishers, Chur, pp 1–30
Schmitt G (1983) Die Comptoise Uhr. Müller, Villingen
Schumpeter JA (1934/1912) The theory of economic development. Harvard University Press, Cambridge
Skinner BF (1966) Operant behavior. In: Honig WK (ed) Operant behavior—areas of research and application. Meredith, New York, pp 12–32
Smith A (1759) The theory of moral sentiments. Millar, Kincaid, and Bell, London, Edinburgh
Smith A (1776) An inquiry into the nature and causes of the wealth of nations. Strahan and Cadell, London
Trivers RL (1971) The evolution of reciprocal altruism. Q Rev Biol 46:35–57
Usher AP (1954) A history of mechanical inventions. Harvard University Press, Cambridge

Veblen TB (1898) Why is economics not an evolutionary science? Q J Econ 12:373–397
Walras L (1954/1874) Elements of pure economics—on the theory of social wealth. Irwin, Homewood
Wedervang F (1965) Development of a population of industrial firms. Universitetsforlaget, Oslo
Winter SG (1964) Economic "natural selection" and the theory of the firm. Yale Econ Ess 4:225–272
Witt U (1985) Economic behavior and biological evolution—some remarks on the sociobiology debate. J Inst Theor Econ 141:365–389
Witt U (1991) Economics, sociobiology, and behavioral psychology on preferences. J Econ Psychol 12/4:557–573
Witt U (1992) Evolutionary concepts in economics. Eastern Econ J 18:405–419
Witt U (1996) Moral norms and rationality within populations: an evolutionary theory. In: Pardo JC, Schneider F (eds) Current issues in public choice. Edward Elgar, Aldershot, pp 237–256
Young RM (1988) Darwin's metaphor: nature's place in Victorian culture. Cambridge University Press, Cambridge

Part II Evolutionary Nonlinear Dynamics

Economic Development in the Arts, Crafts, and Sciences*

TÖNU PUU

Summary. Development studies in economics normally focus quantitative growth in well-defined sets of objects. This is not suitable in the case of the arts, where objects are often described as unique. Qualitative change, by which the set structure itself changes in terms of the defining properties, needs a different kind of analysis. In this chapter, the analogy to the Darwinian development tree is invoked. Characteristics of change are considered within the framework of Lancasterian demand theory, with Thom's catastrophe theory being used to study the bifurcation points in the evolution tree. Increasing specialisation of individual implements and the complexity of the whole are emphasized.

Key words. Economic evolution, Arts and sciences, Development tree, Catastrophe theory, Specialization

1 Introduction

Objects or manifestations of art are often characterized as being unique, in contrast to mass-produced goods and services. The same is also true in science, because any reproduction of scientific ideas necessarily is a consequence of ignorance or even intentional fraud.

This makes theorizing about development in the cultural sector particularly hard, because we are accustomed to measure progress by means of growth in the numbers of well-defined objects. On the other hand, progress, or even just development, in these areas is justly seen as an improvement, or merely a change, in quality rather than in quantity.

Most often we try to measure improved quality by quantity, by a priori defining a hierarchy of classes of basically similar objects, preordered in terms of quality, a Rolls Royce having higher quality than a Fiat, and a PhD having higher quality than

---

Department of Economics, University of Umeå, SE-90187 Umeå, Sweden

* Adapted from "Development: Increasing Complexity," in *Attractors, Bifurcations, and Chaos*, by T. Puu (Springer, Berlin and Heidelberg, 2000), 435–466. With the permission of the publisher.

a college graduate. Then, the average quality of an aggregate can be measured quantitatively, in terms of the relative distribution of objects in these classes. The quality of manpower in a plant can thus be measured by the average length of education, and its quality is regarded as being improved if this average increases.

Education time, as used in that example, makes things unduly simple because it is itself a quantitative measure which can be used in ordering the different classes. In most cases such a possibility is not available, and the classification order is bound to become more or less arbitrary. It is much more difficult to say why a Rolls Royce is superior to a Fiat than why a PhD is superior to a college graduate.

Except for the fact that such classifications tend to be arbitrary, one might wonder how to describe or model a development through which a previously nonexistent object is born. This just adds a new dimension to the space we are studying. Mathematical methods do not favor the modeling of such changes.

2 Böhm-Bawerk and Smith

In economics, many ideas of improvement through qualitative change have been advanced, but none has been translated into a successful formal model. There are two excellent examples: Böhm-Bawerk's theory of roundabout production, and Adam Smith's theory of labor division.

Böhm-Bawerk, in his *Capital und Capitalzins* from 1900, discusses how, step by step, the production process becomes more and more roundabout, but pays off in terms of an increased productivity which more than compensates for the time spent on capital formation:

> A farmer needs and wants water for drinking. The spring bubbles at some distance from his house. In order to satisfy his need for water he can choose different ways. Either, he goes each time to the spring and drinks from his hollowed hand... Or... —the farmer makes a jug from a block of wood and carries in it his daily use at once from the spring to his living place... but in order to achieve this it would have been necessary to make a not insignificant detour: the man might have to carve on this jug a whole day; he should, in order to be able to carve it... fell a tree... even before that produce an axe, and so on... Finally... instead of one tree he could fell many, hollow them all in the centre, build a pipe from them and bring an abundant stream of water to his house. Obviously roundaboutness is now even more considerable. As a compensation it leads to a much enhanced achievement.

Likewise, Adam Smith in his *Wealth of Nations* from 1776, speaks almost lyrically of the productivity increases resulting from specialization of labor. His famous example is that of specialized pin-making, where both organization and training, combined with specialization in accordance with natural talent, result in improved productivity:

> One man draws out the wire, another straights it, a third cuts it, a fourth points it, a fifth grinds it at the top for receiving the head; to make the head requires two or three distinct operations; to put it on, is a peculiar business, to whiten the pins is another; it is even a trade by itself to put them into paper; and the important business of making a pin is, in this manner, divided into about eighteen distinct operations... Each person, therefore, ... might be considered as making four thousand eight hundred pins a day. But if they had all wrought separately and independently, and

without any of them having been educated to this peculiar business, they certainly could not each of them have made twenty, perhaps not one pin in a day.

It is interesting to consider how economists nowadays would model the benefits of roundaboutness or specialization. Roundaboutness would be introduced as an increase of some homogeneous input of capital in the production function, a proxy for the quantity of jugs and pipes, whereas division of labor would just be represented by a multiplicative productivity constant, multiplied either to the quantity of some again homogeneous labor input, or else just to output, and increasing over time. This practice must be regarded as very crude, doing no justice at all to the original ideas. It is obvious that not much of the phenomena discussed by Böhm-Bawerk and Smith is caught in this way.

3 Increasing Complexity

Economic development always has one aspect of increasing mass or volume: of people, households, commodities, services, firms, or communications.

There is another salient feature, a qualitative aspect of increasing complexity, diversity, and specialization of function. Typically, the various components of a more complex whole also seem to become internally more complex—more complex the more specialized their function is. This development towards increasing specialization and internal complexity seems to be almost a natural law, like the degradation of energy in thermodynamics, although it has a reverse direction to that represented by energy degradation, towards higher order and organization!

In this respect, economic development is similar to the development of the species in zoology: from a few primitive organisms at a low level of specialization and internal complexity, to highly organized ecosystems of specialized and complex organisms. It can be maintained that this difference in direction (towards increasing "enthalpy" in stead of "entropy") is a salient difference between living and dead matter. Human society, in this respect, then seems to be just one manifestation of life.

The economics of development, as indicated, focuses on increasing quantity rather than increasing diversity. This is natural, because quantitative increases in well-defined sets of objects are always easier to analyse than increasing diversity of the set structure itself. Yet the latter is the more interesting feature in the development process. It therefore seems necessary to take a different approach to modeling economic development whenever qualitative issues have to be considered, even if that approach may have to be a simplistic one.

4 The Development Tree

Such a starting point could be an analogy to the Darwinian development tree. As can be seen from archeological findings, the development of the tools of primitive man started with multipurpose stone tools, with no differentiation between the operations of cutting, scraping, and sawing, just as zoological development started with primitive organisms of low differentiation, such as amoebas.

In the course of time, the saw developed as distinct from the knife and from the scraper, i.e., the lines branched exactly as they do in the biological development tree. Some reflection convinces us that this process of branching and increasing specialization is universal in the development of whatever human implements we study, be it just simple hand tools, or musical or scientific instruments.

The question is, what is the reason for this increasing specialization? The idea of regarding the development of human implements as analogous to the development of biological organisms is not new. A few contemporaries of Charles Darwin actually saw things this way. Among those was Karl Marx, who, however, found no use for parallels to the ideas of mutations and survival. Accordingly he did not detect the great potential for Darwinian evolutionism in economic development. If development is just a result of wilful planning by humans, for better or for worse, then there remains nothing to model.

However, anthropologist Gordon Childe, in the 1940s, emphasized these deeper parallels, and George Basalla in *The Evolution of Technology*, suggests both mutations and innovations for the sheer fun of *homo ludens*, and a selection mechanism due to various socioeconomic forces. Basalla poses the question whether it would really be likely that such variety as we find in the real world would result from conscious planning to satisfy human needs concerning shelter, nourishment, and the like. Why, for instance, should about 500 different kinds of specialized hammers have been produced in Birmingham in 1867? Moreover, he asks whether the automobile, constructed when the combustion engine was invented by Nikolaus Otto in 1876, really was necessary. He claims that the automobile was essentially a toy in the decade 1895–1905, and that in no way was it triggered by any kind of shortage of horsepower.

An even more basic invention, the wheel, which seems to have been invented in the Near East about 5000 years ago, was in fact completely scrapped in the Middle Ages in that very region, and transportation made use of the more efficient camels rather than carriages drawn by oxen or horses. This state of things remained for a period of about 1000 years! Basalla also cites the remarkable fact that the Aztecs of Mexico used working axles and wheels for animal figurines used for ritual purposes or just as toys, although there is no evidence at all that wheeled vehicles were in use for any practical purpose. Basalla gives good reasons for why this was so: carriages were just not very useful in the jungles and mountains of Mexico. Likewise, with the decay of Roman power in the Near East, one can well imagine that because the need for military transport decreased, the roads were just left to decay.

No matter what the reasons, it is interesting that these things were invented without being at all useful, and hence became more like random mutations in biology, just set out to test whether they were useful for anything (even if only as toys). They might then linger on, and their use in practice might explode when a niche opened up, as in the case of the imported American wine-louse which almost destroyed all European vineyards, or the decorative weed that is about to transform Lake Victoria into a swamp today.

Basalla pursues the analogy between the evolution of living organisms and that of human artifacts in great detail. First of all he compares the number of different species, estimated to 1 500 000, with the number of patents, as representing different "species" of artifacts. Those taken out in the US only from 1790 to around 1980 amounted to 4 700 000. He therefore concludes that the complex richness of the world of artifacts is hardly less pronounced than the world of living organisms.

Of course, not all patents represented economically or even technologically viable inventions. As late as in the period 1855–1905, when the laws of thermodynamics were known, more than 500 patents were granted by the British Patent Office for various perpetual motion machines, and only in 1914 did the US Patent Office issue a general declaration that any drawings for perpetual motion machines had to be accompanied by working models. The number of these patents indicates an order of magnitude. A proportion of 10%–50% of all patents have been judged to have economic significance. The remaining useless fraction is interesting too, as it demonstrates the urge in the human mind to invent things just for the pleasure of it, so justifying the "mutation" analogy.

Earlier writing on these matters, such as Samuel Butler's essay *Darwin and the Machines* dating from 1863, classifies the mechanical "life" in genera, species, and varieties, and attempts to sketch a development tree in an analogy to Darwin's. In this work there is an absurd preoccupation with the lack of reproductive capacity among machines, and attempts to make the analogy complete by considering machines making machines on their own. Coupled with this is a fear that machines somehow would "take over" the role of mankind. At this stage, however, the analogy ceases to be fruitful. We do not need self-reproducing machines because humans, by reinvestment, make sure that viable things are reproduced.

Basalla also claims that there is no clear idea of a selection mechanism in this early work, and that they flip between the idea of technological evolution being a continuous, collectively social process, and that of it being a discrete process of inventions with great names labeling each one of them. He stresses that there is nothing exclusive in the approach, and that the actual historical development in relevant technology displays continuity as well as discontinuity.

However, he does not consider the ideas of mathematical bifurcation, competing attractors, basins of attraction, and niches for the survival of human implements, nor the general way in which the totality of technology feeds back to provide these niches. All this could become much more understandable using such simple mathematical ideas.

5 Continuous Evolution

Basalla gives detailed evidence for the fact that the common impression that technological development is a sequence of discrete revolutionary inventions is in part based on an inadequate knowledge of the actual history of technology, which makes us ignorant of all the minor stepping stones between the great historical moments.

This is similar to the case of reading a concise history of music, focusing on, say, Palestrina, Monteverdi, Bach, Mozart, Beethoven, Brahms, and Schönberg. The discrete steps are giant. However, considering the music by Bach's sons and their contemporaries of the "gallant style" makes the jump to Mozart much more understandable. It makes it easier to isolate his real contribution, without subtracting in the least from his genius. In the same way, intermediate styles fit in between any of the named giants, thus making the whole evolution process for composition styles a much more continuous process.

Basalla takes as his starting point James Watt's invention of the steam engine in 1775, in the romanticized popular representation where it is shown as the result of a dream by a boy inspired by watching a boiling tea kettle. Basalla points out the preexistence of blacksmith Thomas Newcomen's steam engine for pumping water out of mines, already invented in 1712. Although the principles are quite different (Newcomen's engine worked with a single action in a vacuum resulting from the condensation of cooled steam), it is a fact that Watt's invention was a result of the repair and improvement of one of the Newcomen engines at the University of Glasgow, where Watt worked as an instrument maker. Basalla also points to the fact that not even Newcomen's invention appeared out of nowhere, all the elements, such as piston pumps, mechanical linkages, and steam displacement devices, being well known and used in Europe in the 17th century. In fact, Thomas Savery's invention the Miner's Friend, demonstrated in 1699, even antedates Newcomen's machine.

He also points out links to the later internal combustion engine, since in 1759 it had already been proposed that steam should be replaced by hot air in steam engines, and in 1791 an internal combustion "steam engine," working on turpentine vapor, was patented. Although the two-way action from the original steam engine still remained, nevertheless this can be seen as a stepping stone toward the giant leap to Nikolaus Otto's combustion engine.

More surprising is that even early electric motors were modeled directly after the steam engine, with pistons moving back and forth in cylindrical electromagnetic coils. The rotary motion we take as a matter of course in the electric motor (although not necessary, as this example shows) also has its obvious mechanical analog in the steam turbine, which itself, of course, goes back to very old ideas embodied in water wheels in mills and the like.

All this demonstrates the usefulness of the biological analogy. First, there no doubt exists a continuous evolution process in the realm of human artifacts as in that of living organisms. Second, there is such an overwhelming richness of "species," that it must have arisen through a process of branching and diversification, rather than by some act of simultaneous "creation," which is the obvious alternative.

6 Diversification

Let us try to pin down some facts about the development of human artifacts, keeping in mind the bifurcation idea. We may now observe something like a curious natural law: given any state of technology, an implement can be made more efficient the more specific its operation is. The original all-purpose tool of the Stone Age can never be as efficient as a saw for sawing, or a knife for cutting. On the other hand, an all-purpose tool is better for sawing than is a smoothly sharpened knife.

As a consequence, in the process of harvesting increased efficiency by specialization, something of the versatility and efficiency of secondary functions is sacrificed. Only when the different specialized tools are again assembled in a joiner's workshop, or in a modern combination machine, is there an overall improvement in efficiency.

It is very unlikely that such a development would, in general, proceed according to some premeditated blueprint. Rather, we would expect that a more knife-like tool is developed just when efficient cutting is desperately needed, and the saw-like tool when sawing is particularly essential (in terms of economics; when the prices for these services are high). The complete workshop or combination machine is likely to be organized first when most of the specialized tools are already at hand. This will then be the result of recognizing the complementarity properties of various specialized tools when they are used in combination.

This is also similar to the case of zoology. Ants and anteaters are obviously complementary from an ecological point of view, even though an individual ant, provided it had a conscious mind, might be reluctant to distinguish the anteater as its natural companion.

Such increased complexity can develop at several levels simultaneously. The joiner's workshop is an example at a relatively low level; the making of furniture in the 18th century, involving a multitude of different workshops and craftsmen of different specializations, among them the joiner, is another—at a higher level.

The reader may consult Edward Lucie Smith 1981, *The Story of Craft—The Craftsman's Role in Society*, or the *Dictionnaire raisonnée des sciences, des arts et des métiers*, by d'Alambert and Diderot, about this subject.

7 Property Space

In mentioning the properties of a tool, exemplified by the ability to cut, to saw, or to scrape, it was implicit that tools and other human implements could be regarded as bundles of properties.

To make things more specific, we can treat them as measurable efficiencies in various operations, thus allowing the implements to be represented as coordinate points in some space of sufficiently many dimensions. This is the way consumer goods were once regarded by Lancaster (1971) in his seminal contribution *Consumer Demand. A New Approach*. It seems fairly obvious that the performance of man-made implements, as discussed above, exactly like the case of living organisms, can, for any practical or scientific purpose, be represented as points in a Lancasterian property space of sufficiently many dimensions.

Some dimensions might be hard to quantify for cultural objects, but the difficulties seem to be far from insurmountable. In fact, critics of art, literature, and music have a habit of discussing the quality aspects of a given product in different dimensions separately before amalgamating them into a compound evaluation. Experiments show that such critics are remarkably unanimous when listing and characterizing the important aspects of a work of art, even if their judgment in individual cases differs.

Exactly the same holds true in evaluations of scientific work judged in terms of originality, analytical acumen, clarity of presentation, practical usefulness, and the like. This is done constantly in the evaluation of staff for promotion, or in the distribution of research funds among applicants. As is done in many other scientific procedures, it is then deemed to be reasonable to introduce some scale for measuring the score in each dimension. This is, in fact, done by every teacher grading exams.

As the next step, we would place the Darwinian development tree in a Lancasterian property space. This is a new setting for the development tree; the branching tree structure remains, but it is placed in a multidimensional vector space. The dimensions of the space in which the biologists usually draw the development tree do not make such mathematical sense; it is just a two-dimensional picture of a set of three-dimensional animals. In this respect our use of the analogy is different.

The essential characteristic of the development tree is its combination of continuous development and branching structure. By the branchings or bifurcations, we can best understand the nature of the emergence of new implements with more specialized functions.

8 Paretian Ordering

Let us turn to the issue of evaluation. Within the setting of Lancasterian property space, economics offers just one uncontroversial way of defining progress: the Paretian.

Whenever one tool performs better in at least one dimension without performing worse in any other, for instance being better as a knife without being worse as a saw or a scraper, then we would say that it is better in a Paretian sense. As indicated above, natural limits seem to favor efficiency in a specialized tool. Therefore, in the Paretian meaning just defined, we very seldom encounter a case of real progress. Most changes in history represent improvements in some dimensions at the expense of others.

The chance for such Paretian comparability of two objects, chosen at random in a group, increases with the number of objects in the group, and decreases with the dimensions of the Lancasterian space in which they are represented. We can even present a simple formal argument for the latter. In order to retain a certain degree of Paretian comparability, the number of objects has to increase geometrically when the dimension of the space increases arithmetically. The reason is that a point in n-dimensional property space divides the space into 2^n orthants, among which only two are comparable in Paretian terms. Consequently, incomparability increases very fast in relation to the dimensions of the Lancasterian property space.

Very complex mass-produced goods, of which there are many examples in our high-tech society, may also need a property space of many dimensions for their representation, and comparisons of different brands have basically the same characteristics as comparisons of different works of art. Nevertheless, as long as there are many copies of each design, it still makes some sense to define one dimension for each brand, and to focus analysis on their numbers. In the case of artwork, this is not meaningful, as there may be just one copy of each design.

In practice, we expect the number of implements in every special group to more or less equal the number of dimensions in the property space. This is due to the possibility of combining them, just like the tools in a joiner's workshop. If the efficiency of the combined performance were linear, we would even expect a strict equality to hold.

Too many tools would then imply that some of them were redundant, since they were dominated by a convex combination of the remaining ones; too few would make it possible to span only a subspace of the space of operations. Of course, we cannot expect things to be as simple as that. In particular, we have to consider that we are dealing with capital goods with long technological life-times, so at any moment we should expect to see vintages of tools which do not represent the latest state of technology.

9 Prices and Progress

Thinking in terms of economics, we could, however, always get a step further by evaluating heterogeneous bundles of properties with the help of a price system.

The question is how we are to obtain such prices. A natural way to proceed would be to calculate the imputed prices for the various properties from the actual market values of the commodities possessing them in various degrees. We could then calculate the values of a given tool in different designs. According to the theory of production, we would expect a change of design to occur whenever prices change, and the value would then be higher for the new design as compared with the old at every transition, because otherwise it would not have occurred.

It would also be understandable that major changes in design need not be the results of inventions and discoveries; they may just result from changing evaluations, reflecting changing needs, tastes, habits, and whims. History can even be circular, and return to an old solution.

Paradoxically, the value of an object in the course of changing design, always calculated at current prices, would show up as constantly increasing, even when the development is circular. Economic index theory then warns us that there is a problem here. One can evaluate each single design change in the history of a certain human implement in terms of the momentary set of prices, and report an increase in value, but one cannot just accept a continuous increase in value when new prices are applied at each transition. This, however, is exactly what historians reporting constant progress do.

It is interesting to note that historians of science, art, and music of various ages usually report continuous progress until their own times, when some recent transition renders a state of final perfection. Some later observer may report yet another development and a new state of perfection, but they often accept the previous history. This is very like the attitude of biologists who regard the development of living organisms as a unidirectional way to higher levels of performance, whether the reason is ascribed to divine guidance or to natural selection.

Our claim is that most often something is sacrificed in the apparently constant process of "improvement" in single implements. According to what was said above, improvements in certain dimensions are often bought at the expense of poorer performance in others.

10 Branching Points

We understand that a new direction in the design of implements is most often due to changes in needs and tastes. Such changes on the demand side must be complemented by corresponding design changes on the supply side. Design ideas occur quite frequently, and even erratically, over time. Some of them become successful, some are skipped from the beginning. There are two issues involved here. First, the innovation must be transformed from a blueprint idea into a material implement, and second, it must remain at the stage of being useful over it lifetime, and even be reproduced when it is worn out.

Again, taking the development tree analogy as support, we would say that they arise like mutations in biology. Exactly like mutations, they usually do not remain permanent. Only when there is a macroenvironment with a suitable niche for them, do they appear and remain. The two stages are the same also. First, the genetic change in a chromosome must result in an individual organism that can live. Second, it must survive in terms of being propagated to further generations.

Such permanence for implements can result from a recognition of the complementarity of specialized tools, and from the potential overall productivity increase from using them in combination. The rise of such a suitable macroenvironment is thus responsible for the actual emergence of new implements. The economic identification of such an environment can always be expressed in terms of prices.

The emergence of new implements is also a matter of transgressing certain critical thresholds, and the nature of the branching points as sudden dramatic events contrasting with smooth continuous evolution becomes understandable. For instance, it is well known that the aeroplane was invented many times, but it became a reality only when the aerofoil wing could be combined with an engine of a sufficient power and a trunk of sufficient strength in relation to its weight. Likewise, a computer of any capacity has been a possibility ever since the days of Pascal. If the space required for the machine, the cost of materials needed, and the time for computation had not been a constraint, a supercomputer of any capacity (although admittedly very slow by our standards) could have become a reality long ago. The case of the computer represents an issue of economic feasibility, the case of the aircraft one of technical feasibility, but in essence they are alike.

Modern systems theory provides tools for analysing branching in terms of the vanishing, emergence, fusion, and splitting of attracting points for such systems. The methods are found under the headings of catastrophe theory, bifurcation theory, synergetics, self-organization, and complexity theory.

The drastic disappearance of previous implements are not difficult to find in the fields of either the arts or the sciences. The same holds true for entire systems; the medieval guild system has disappeared just as completely as the world of the dinosaurs.

11 Music Automata

As an example of how a group of implements can disappear, we take the case of music-making automata. They have a surprisingly old history. For instance, an automatic virginal was listed in the inventory of the possessions of Henry VIII made upon his death 1547. Henry, by the way, was himself a fine musician, who even left a few compositions.

This type of instruments stayed around until about 1930, when they suddenly vanished from the world stage. Automatic operation was applied to keyboard, wind, and string instruments. Recordings of the pieces were first made on expensive brass cylinders with pegs, which were later replaced by punched metal discs, and eventually by paper scrolls. This last technology made live recordings possible quite a long time ago, and several fine composers, including Händel and Mozart, produced music for automata.

There exist punched paper scrolls which are believed to have been recorded by Georg Friedrich Händel and produced for automatic organs. Knowing the speed of the feeding machines, they are useful for determining Händel's own tempi. It is not surprising that Händel, being a cunning businessman all his life, would have tried this way of earning some money. The determination of tempi is not without interest, because historical records indicate that his own performance of the "Messiah" took 45 min less than the shortest recording in existence today.

Such automatic pianos and organs existed until their niche disappeared owing to the advances in the gramophone industry. The most fantastic pieces, produced just before the collapse of the whole genre, are machines where an automatic piano has a cupboard above it containing several violins mounted within the circle of a constantly turning resined bow. While the piano is played by the invisible hands of the pianist, the violins are moved toward the circular bow, either one by one or several at once, at any of four different angles, to allow the right string to be touched, while various positions on the strings are pressed by mechanical fingers operated through pneumatic tubes.

From approximately the same period came integrator frames, looking like those simple pantographs which children used to enlarge drawings, although provided with a few gears. By letting the point traverse any closed plane curve, no matter how complex, its three dials computed the perimeter, the area enclosed, and the moment of inertia with respect to some given point in one single step.

The mathematical instruments invented included not only a variety of tools for specific purposes in engineering and physics, and for practical applications in industry or ballistics, but also general purpose instruments such as the Bush (1928) and Weiss (1944) integrators for differential equations, and the mechanical harmonic analyzer devised by Lord Kelvin (1878).

Of course, because the expense of these implements was considerable, they disappeared because they could not compete with the CD or with modern PC software.

The author of this chapter once invented a nomogram for the calculation of internal rates of interest for bonds from nominal rates, prices, and durations. Compared with the enormous tables then used in banks, the cost was reduced to a vanishingly tiny fraction, but alas, in a few years every pocket calculator was supplied with a program for financial computations, so there was no money in it. This was a mutation that did not even survive infancy.

12 Bifurcations

Let us now turn to mathematical tools for the analysis of bifurcations in the development tree we have in mind. Earlier models for dynamic systems were focused on smooth movement only. Nowadays, the focus is more and more on sudden transitions. René Thom, in his 1972 monograph *Stabilité structurelle et morphogénèse*, opened up what used to be called catastrophe theory. His intention was to formulate a method by which structural change, such as the development of a bud into a leaf or a flower, could be modeled mathematically. For a while, this was a most active research area in applied mathematics, and all sorts of phenomena, such as stock market crashes, attacking dogs, riot outbursts, and anorexia/bulimia nervosa, were discussed by these means. Somewhat later, more general theories of "nonequilibrium phase transitions" and "synergetic self-organization" were created on this basis by the double foci of Ilya Prigogine in Brussels and Hermann Haken in Stuttgart.

Today, catastrophe theory is subsumed under the more general topic of bifurcation theory, which is a subject of intense study in connection with deterministic chaos. The latter applies to gradient systems, i.e., those that optimize something, whereas the former applies to dynamic systems in general, whether they optimize anything or not.

If we assume that economic development is the result of some kind of optimization, subject to a changing macroenvironment in terms of changing prices and general know-how, we can restrict our interest to catastrophe theory. This theory has reached a long way in terms of classifying the kinds of bifurcation phenomena that can occur in dynamic systems, depending on the dimension of their state space, and on the number of parameters involved.

In our case, the state space would be the Lancasterian property space, and the parameters would be represented by such things as innovations changing the state of technological knowledge, and, above all, the imputed prices that reflect current tastes. Under certain parameter changes, a previously stable optimum, which for a long time had moved slowly in concordance with changing parameters, might lose stability and jump to any of the new optima into which the previous one bifurcates. Assuming that several agents operate in the same system, they may together recover several or all of the new stable optima, and thus create the branching tree. This explains bifurcations of the type we have been talking about.

A word should be added here about the "production possibility surface," which is embedded in the Lancaster space, and on which continuous change as well as sudden bifurcations occur. It is not of the smooth type found in textbooks on production theory. We have already pointed out that technological oppurtunities seem to favour specialization of function. This means that the possibility surface is concave rather than convex, as is normally assumed. Further, it has a patchy appearance, and so is not even made in one piece. (If it were, we would see cars that are dishwashers and computers at the same time.) This, however, is not necessarily a contradiction of traditional production theory. For one thing, production theory concerns the space of commodities rather than the space of properties, and for another, convex combinations of operations are taken as given in the very definition of a production function, so any concavities are just defined away.

Two words of warning are needed if we want to make explicit models using catastrophe theory. First, we assumed that the Lancasterian property space would need to have a considerable number of state variables, whereas all catastrophe theory attained was a classification of systems with one or two state variables and no more than five parameters. Second, even though the mathematical proofs that all possible phenomena can be modeled in terms of the "canonical forms" of the catastrophes and their "universal unfoldings" are safe, it is actually extremely hard to transfer substantial models into these canonical forms. This is why so many trivial catastrophe models were created.

13 Synergetics

Regarding the first argument concerning the multitude of variables, the Haken's synergetics school provides a good argument why those low-dimensional problems are not so bad after all (Haken 1983). In reality, despite the multitude of variables involved, the probability that more than a few of them are involved in actual phase transitions is very small. The stability of a certain variable is measured by the so-called "eigenvalue," which is obtained by linearizing the system in the neighborhood of an equilibrium. This can be zero, negative, or positive. Positive eigenvalues signify instability, and hence they can not exist for more than short periods otherwise the system will blow up. Negative eigenvalues mean stability, and in the perspective of what eigenvalues close to zero can cause, they can be regarded as momentarily damped out to equilibrium. The interesting things in a system, i.e., catastrophes and bifurcations, are triggered by those few eigenvalues that are very close to zero and that for a while can become positive.

From probability considerations, it is very unlikely that more than a few of all the negative eigenvalues are close to zero. The philosophical argument is that this movement of the very few "slow" variables is the reason why we find a relatively large degree of order and relatively few phase transitions in reality. As a result, low-dimensional models may still be useful for high-dimensional systems.

However, again we need a word of warning. There is no guarantee whatever that those variables associated with the eigenvalues are the variables we measure in reality or use in modeling. They may well represent abstract dimensions constructed in the mathematical normalization process. Hence, we can not expect to come across fast and slow variables among those "natural" ones which we use in modeling.

14 Viable Alternatives

It is obvious that seeing development in this way as a branching process, resulting from more or less erratic mutation-like innovations which, if viable, result in a branching of the development tree, means that the actual state of the world and the route by which it was reached is by no means a unique possible outcome.

It is quite possible that development could have taken another course, so that we might have been equipped with quite different sets of implements, constituting quite different macroenvironments.

Basalla mentions the interesting fact that all Oriental hand tools, be it saws or planes, operate on the motion of drawing the tool toward you rather than on the motion of pushing it away from you. According to Basalla, the Oriental way is even more efficient than the Occidental way. The interesting thing, however, is that both work when they are used in the correct way, and when the labor force is used to them.

Brian Arthur, of the Santa Fe Institute, has focused on such issues as how the QWERTY layout of typewriter keys became established through people just becoming accustomed to it and wanting to have it retained. The situation may thus be static for a long time, although there is no compelling advantage in such a system. On the contrary, he claims that the layout was devised in 1873 with the aim of actually slowing typists down, in order avoid jamming the keys of the old mechanically slow typewriters. Yet QWERTY is still there on my computer keyboard this very moment. So there definitely are possible alternative development histories for human implements, as there are for living organisms.

15 Conclusions

We now summarize what we have concluded about the development process seen from a qualitative point of view. The reader can find a formal analysis along the lines outlined here in my recent book *Attractors, Bifurcations, and Chaos—Nonlinear Phenomena in Economics.*

Proposition 1. Economic development, along with its characteristic of growth, is in the direction of increasing the functional specialization of its elements. The internal complexity of these elements in a more complex whole also increases with the specialization of their function.
Proposition 2. In any given state of technology, an implement can be made more efficient the more specialized is its use. Varying prices for these specialized operations call forth all, or at least most, of the specialized implements. This explains the increase towards specialization.
Proposition 3. The complementarity property of a set of specialized implements when they are used in combination enhances overall efficiency and explains the development towards increasing structural complexity.
Proposition 4. The performance of the implements can be represented as a point in a Lancasterian property space of sufficiently many dimensions.
Proposition 5. The development of implements can be understood by the branching structure of a Darwinian development tree put in the strict framework of a Lancasterian property space.
Proposition 6. In a Lancasterian property space, Paretian comparability is very unlikely to occur, especially if the space has many dimensions. Therefore, we seldom see indubitable progress in culture.
Proposition 7. The preferences of the economy are reflected in the imputed prices for the various performance dimensions of the Lancasterian space. When commentators speak of progress, they mean increasing value in terms of the imputed prices ruling at that moment.
Proposition 8. Innovations come like mutations at a constant but erratic rate. They result in technological change when a niche is opened up for them by a suitable macroenvironment, reflected in favorable imputed prices.
Proposition 9. Mathematically, the branching points in the development tree can be analyzed by bifurcation or catastrophe theory.
Proposition 10. Ideas developed by Haken's synergetics school explain how the structural change in the branching tree can be low-dimensional even though the whole system is obviously high-dimensional.

References

Arthur B (1990) Positive feedback in the Economy. Sci Am February:92–99
Basalla G (1988) The evolution of Technology. Cambridge University Press, Cambridge
Böhm-Bawerk E von (1900) Capital und Capitalzins. Verlag der Wagner 'schen Universitäts tuchhandlung, Iurns Ructs
D'Alambert J le Rond, Diderot D (1751) Encyclopédie, ou Dictionnaire raisonné des sciences, des arts et des métiers. Paris
Haken H (1983) Advanced synergetics. Springer, Berlin etc.
Lancaster K (1971) Consumer demand. A new approach. Columbia University Press, New York
Lucie Smith E (1981) The story of craft—the craftsman's role in society. Phaidon, Oxford
Puu T (2000) Attractors, bifurcations, and chaos—nonlinear phenomena in economics. Springer, Berlin etc.
Schumpeter J (1954) History of economic analysis. Allen & Unwin, London
Smith A (1937) The Wealth of Nations. Random House, New York

Nonlinear Dynamics of Debt and Capital: A Post-Keynesian Analysis

Toichiro Asada

Summary. We reconsider the dynamic theory of debt and capital from a post-Keynesian perspective. First, we present a microeconomic model of the debt effect on investment activity which is based on the "principle of increasing risk" in the manner of Kalecki. Then a simple macrodynamic model of debt and capital accumulation, which produces the "Minsky cycle," is developed. Finally, we briefly discuss the macroeconomic implications of the instability of a flexprice economy with a debt effect.

Key words. Debt accumulation, Capital accumulation, Post-Keynesian approach, Principle of increasing risk, Minsky cycle

1 Introduction

In traditional versions of macroeconomics, the impact of debt on macroeconomic performance is almost ignored. This statement applies to the standard textbook version of "Keynesian" macroeconomics as well as the tradition of "classical" or "new classical" macroeconomics. For example, in the standard version of the IS-LM model, we cannot find such a variable as debt. Without doubt, this ignorance of debt contradicts the widespread experience of Japanese, Asian, and US economies in the 1980s and 1990s. However, there is another (heretical) tradition in economic thought which stresses the importance of the impact of debt on macroeconomic stability or instability. The most famous examples are Keynes himself (1936) and Fisher (1933).[1] In modern economics, the "post-Keynesian" school inherits this vision and stresses the importance of debt. The most influential scholar in this line of thinking is Hyman P. Minsky.[2] Minsky himself did not develop fully fledged mathematical models, but recently various formal models which purport to capture the essence of Minsky's ideas have been developed

Faculty of Economics, Chuo University, Higashinakano, Hachioji, Tokyo 192-0393, Japan

[1] Keynes (1936) writes as follows: "Indeed if the fall of wages and prices goes far, the embarrassment of those entrepreneurs who are heavily indebted may soon reach the point of insolvency—with severely adverse effects on investment. Moreover the effect of the lower price-level on the real burden of National Debt and hence on taxation is likely to prove very adverse to business confidence." (p. 264)

[2] See, for example, Minsky (1975, 1986).

by some economists in the post-Keynesian tradition.[3] Their contributions are essentially based on the recent developments of nonlinear dynamical economics, because the problem is intrinsically dynamic in the nature of things, and nonlinear dynamical economics offers the powerful tools needed to present their vision. Economic life is full of complexity which originates from nonlinearity. Nonlinear dynamical economics can offer an interpretation of the complex behavior of such economic variables as debt or capital by referring to only a few relatively simple principles.[4]

In this paper, we reconsider the dynamic theory of debt and capital from a post-Keynesian perspective. We begin our discussion with consideration of the microeconomic foundation of the investment function with a debt effect. We argue that the "principle of increasing risk," which was introduced by Kalecki (1937), who is a pioneer of post-Keynesian economics, gives a sound basis for the microeconomic foundation of such an investment function. In Sect. 3, we develop a simple macrodynamic model of debt and capital accumulation. The model presented is formally similar to the Volterra–Lotka system of mathematical biology, which is designed to describe the struggle between predator and prey, and it produces the so-called "Minsky cycle". Finally, in Sect. 4, we briefly discuss the macroeconomic implications of the instability of a flexprice economy with a debt effect. It will be argued that an increase in price flexibility may contribute to destabilizing rather than stabilizing the economy, which is contrary to the traditional teaching if the debt effect is strong.

2 Microeconomic Foundation of the Investment Function with a Debt Effect

In traditional investment theories, including Keynes' own theory as well as neoclassical theory, the financial variables such as debt are disregarded. This is the inevitable consequence of an assumption of a perfect capital market rather than a logical flaw. In a world with a perfect capital market there are no financial constraints, and a firm can borrow without restriction at the given market rate of interest. It is well known that in such a situation the financial decisions of the firm are irrelevant to the market value of the firm, and the optimal capital structure is indeterminate.[5] However, the recent development of "new Keynesian" as well as "post-Keynesian" modeling of macroeconomic fluctuation stresses the importance of the interaction between investment and the financial decisions of firms in a world with an imperfect capital market.[6] In particular, the "financing hierarchy view" (or "financing pecking order view") of corporate finance, which was developed by Baumol et al. (1970) and Fazzari et al.

[3] The following list represents examples of recent contributions: Arena and Raybaut (1998); Delli Gatti and Gallegati (1994, 1995); Franke and Semmler (1989); Jarsulic (1990); Keen (2000); Kuroki (1994); Pally (1996); Taylor and O'Connell (1985).

[4] For surveys of nonlinear dynamical economics, see, for example, Asada (1997), Flaschel et al. (1997), and Lorenz (1993).

[5] This is the famous conclusion which is derived from the so-called "Modigliani–Miller theorem" (cf. Modigliani and Miller, 1958).

[6] For "new Keynesian" literature, see, for example, Bernanke and Gertler (1989), Fazzari et al. (1988), and Greenwald and Stiglitz (1993).

(1988), is important as a "new view" which was developed in this new tradition. According to this view, the cost of external finance is higher than the cost of internal finance in a world with an imperfect capital market with default risk and assymmetric information. In such a situation, external finance and internal finance are not interchangeable, and a firm will resort to external finance only when the available level of internal finance is lower than the level of investment expenditure.

We now pose the following question: "How new is the above mentioned 'new view' of corporate finance?" The answer is that it is more than 60 years old. In fact, we can trace the origin of such a view to the "principle of increasing risk" which was presented by Kalecki (1937) a year after Keynes' "General Theory" was published.[7]

Kalecki (1937) presented a modified criterion of optimal investment which is an alternative to Keynes' (1936) famous criterion ($m = \rho$). This is

$$m = \rho + \sigma \tag{1}$$

where m is the "marginal efficiency of investment," ρ is the market rate of interest, and σ is the "marginal risk," which is assumed to be an increasing function of the amount of investment expenditure.

Asada and Semmler (1995) and Asada (1999) showed that the Kaleckian approach to "increasing risk" can be considered as a promising microeconomic foundation of the post-Keynesian or new Keynesian theory of investment and finance with an imperfect capital market. Next, we interpret the essence of the idea by using the simplest version of the model, which was presented by Asada (1999).

We consider the effect of investment $I = \Delta K$ under the assumption that the investment is done only in the initial period (period 0), and that the expectations of an entrepreneur about the rate of profit and the rate of interest are static. In this case, we can express the discounted present value of the net cash flow (A), which is to be maximized, as described below.[8]

$$\begin{aligned} A &= \int_0^\infty r(K+\Delta K)e^{-\rho t}\mathrm{d}t - \phi_1(g)K - \int_0^\infty \phi_2((D+\Delta D)/K)Ke^{-\rho t}\mathrm{d}t \\ &= \int_0^\infty r(1+g)Ke^{-\rho t}\mathrm{d}t - \phi_1(g)K - \int_0^\infty \phi_2((D+\Delta D)/K)Ke^{-\rho t}\mathrm{d}t \end{aligned} \tag{2}$$

where r is the rate of profit, ρ is the rate of discount, which is assumed to be the market rate of interest of the safe asset, K is the real capital stock, $g = I/K = \Delta K/K$ is the rate of investment (rate of capital accumulation), D is the initial level of debt, and ΔD is the increment of debt which is due to the investment activity in the initial period.

We assume the functional forms of the functions ϕ_1 and ϕ_2 to be

$$\phi_1(0) = 0, \qquad \phi_1'(g) > 0, \qquad \phi_1''(g) > 0, \qquad \phi_1'(0) = 1 \tag{3}$$

$$\phi_2(0) = 0, \qquad \phi_2'((D+\Delta D)/K) > 0, \qquad \phi_2''((D+\Delta D)/K) > 0 \tag{4}$$

for $(D + \Delta D)/K \geqslant 0$.

---

[7] For the development of this line of thought, see Steindl (1945), Steigum (1983), Asada and Semmler (1995), and Asada (1999). See also Kalecki (1971) Chap. 9.

[8] We normalize $p_k = 1$ without loss of generality, where p_k is the price of the capital good.

The function ϕ_1 captures the hypothesis of the increasing adjustment cost of investment which was introduced by Uzawa (1969). This hypothesis of increasing adjustment cost was called the "Penrose effect" by Uzawa. On the other hand, the function ϕ_2 captures, so to speak, the "Kalecki effect," which implies that marginal cost as well as total cost of default risk increases as the stock of debt relative to the capital stock increases.

If, for simplicity, we assume that there is no issue of new shares, the budget constraint of the firm becomes

$$\Delta D = \phi_1(g)K - s_f(rK - iD) \tag{5}$$

where i is the rate of interest which is applied to the initial debt, and s_f is the rate of internal retention ($0 \le s_f \le 1$). Usually, i will be higher than ρ. We assume that

$$i = \rho + \varphi(d); \qquad \varphi(d) \ge 0, \qquad \varphi'(d) \ge 0 \tag{6}$$

$$r - id = r - (\rho + \varphi(d))d > 0 \tag{7}$$

for the relevant range of d, where $d \equiv D/K$ is the debt–capital ratio. (We neglect the repayment of debt for simplicity.) We can then rewrite Eq. 5 as

$$(D + \Delta D)/K = d + \phi_1(g) - s_f\{r - (\rho + \varphi(d))d\} \tag{8}$$

The optimal program for the firm is given by the combination (g, s_f), which maximizes A under the constraint which is expressed by Eq. 8. Asada (1999) derived the following set of optimal conditions.

$$s_f = \bar{s}_f \tag{9}$$

$$r/\phi_1'(g) = \rho + \phi_2'(\cdot) \tag{10}$$

where $\bar{s}_f \in (0, 1]$ is the institutionally given upper bound of the rate of internal retention.

Equation 9 implies that the optimal retention rate is determined as a "corner solution." Obviously, this result is consistent with the "financing hierarchy view" of corporate finance which was stressed by Baumol et al. (1970) and Fazzari et al. (1988). On the other hand, we can consider that Eq. 10 is in fact equivalent to Kalecki's formula, which is expressed in Eq. 1.

First, we can prove that the left-hand side of Eq. 10 is equivalent to Keynes' concept of "marginal efficiency of investment" (m).[9] On the other hand, $\phi_2'(\cdot)$ corresponds to Kalecki's concept of "marginal risk" (σ). Therefore, Eq. 10 is equivalent to Eq. 1.[10]

---

[9] The investment expenditure of the firm, including adjustment cost, is given by $E = \phi_1(g)K$, so that we have $\Delta E = \phi_1'(g)K\Delta g$. On the other hand, Keynes' concept of the "marginal efficiency of investment (m)" is given by the formula

$$\Delta E = \int_0^\infty r\Delta I e^{-mt}\mathrm{d}t = \int_0^\infty rK\Delta g e^{-mt}\mathrm{d}t$$

From these relationships, we obtain $m = r/\phi_1'(g)$.

[10] Asada (1999) also gives an alternative interpretation of Eq. 10 which is based on Tobin's q theory.

We also have the relationships

$$m = r/\phi_1'(g) = m(g; r); \quad m(0, r) = r, \quad \partial m/\partial g < 0, \quad \partial m/\partial r > 0 \tag{11}$$

$$\sigma = \phi_2'(\cdot) = \sigma(g; r, \rho, \bar{s}_f, d); \quad \partial\sigma/\partial g = \phi_2''(\cdot)\phi_1'(g) > 0$$

$$\partial\sigma/\partial r = -\phi_2''(\cdot)\bar{s}_f < 0, \quad \partial\sigma/\partial\rho = \phi_2''(\cdot)\bar{s}_f d > 0$$

$$\partial\sigma/\partial\bar{s}_f = -\phi_2''(\cdot)\{r - (\rho + \varphi(d))d\} < 0$$

$$\partial\sigma/\partial d = \phi_2''(\cdot)\{1 + \bar{s}_f(\rho + \varphi(d)) + \bar{s}_f\varphi'(d)d\} > 0 \tag{12}$$

Equation 12 shows that the marginal risk is an increasing function of the rate of investment, which is nothing but the Kaleckian principle of increasing risk, but the marginal risk also depends on some other variables such as rate of profit, rate of interest, debt–capital ratio, etc. Substituting Eqs. 9, 11, and 12 into Eq. 10, we have

$$m(g; r) = \rho + \sigma(g; r, \rho, \bar{s}_f, d) \tag{13}$$

Solving Eq. 13 with respect to g, we obtain the following investment function with a debt effect:

$$g = g(r, \rho, \bar{s}_f, d); \quad \partial g/\partial r > 0, \quad \partial g/\partial\rho < 0, \quad \partial g/\partial\bar{s}_f > 0, \quad \partial g/\partial d < 0 \tag{14}$$

We can interpret the determination of investment in our model using Fig. 1.

If $\sigma = 0$, the optimal rate of investment is determined by the point e in Fig. 1, which corresponds to $\tilde{g}$, and $\tilde{g}$ is independent of the financial variables such as $\bar{s}_f$ or d. However, under the Kaleckian hypothesis of increasing risk, the optimal investment is determined by the point E in Fig. 1, which corresponds to g^*. It is important to note that g^* is smaller than $\tilde{g}$, and that g^* is influenced by financial variables. For example, an increase in d induces the upward shift of the curve $\rho + \sigma$ in Fig. 1, so that g^* decreases correspondingly. This proves that the Kaleckian principle of increasing risk

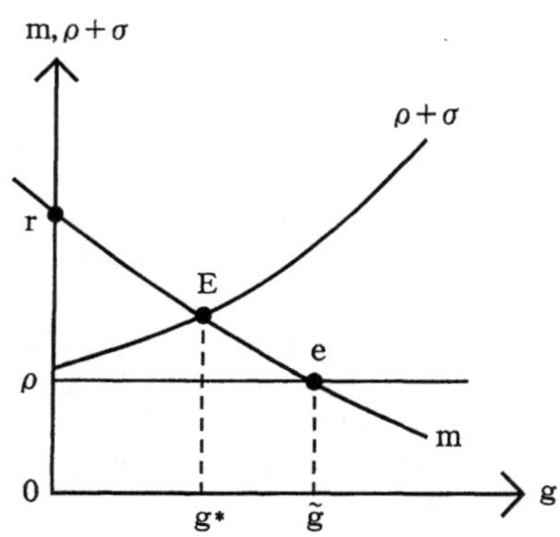

FIG. 1. Determination of investment

can offer a sound microeconomic foundation for the post-Keynesian theory of financial dynamics.[11]

3 Minsky Cycle as a Predator–Prey Struggle: A Long-Run Dynamics of Debt and Capital

In this section, we investigate the macroeconomic implications of the post-Keynesian approach, based on the investment function in the previous section, by constructing a simple model which describes the long-run dynamics of debt and capital.

First, we interpret Eq. 5 in the previous section as the differential equation

$$\dot{D} = \phi_1(g)K - s_f(rK - iD) \tag{15}$$

Next, by definition, we have

$$\dot{d}/d = \dot{D}/D - \dot{K}/K = \dot{D}/D - g \tag{16}$$

Substituting Eq. 15 into Eq. 16, we have

$$\dot{d} = \phi_1(g) - s_f(r - id) - gd \tag{17}$$

Substituting Eqs. 6, 7, and 14 into Eq. 17, we obtain the relationship

$$\dot{d} = \phi_1(g(r, \rho, s_f, d)) - s_f\{r - i(\rho, d)d\} - g(r, \rho, s_f, d)d \tag{18}$$

where $i(\rho, d) \equiv \rho + \varphi(d)$. Equation 18 gives the dynamics of the debt–capital ratio.

We now consider the Keynesian quantity adjustment process in the goods market. We adopt the formulation

$$\dot{y} = \alpha(c + h - y); \qquad \alpha > 0 \tag{19}$$

$$y \equiv Y/K, \qquad c \equiv C/K, \qquad h \equiv E/K = \phi_1(g) \tag{20}$$

where Y is the output level (national income), C is the consumption expenditure, and $E = \phi_1(g)K$ is the investment expenditure including adjustment cost.[12] In this simplified model, we abstract from the changes in price so that there is no distinction between real and nominal values.[13]

For consumption expenditure, we assume that

$$C = C_w + C_r \tag{21}$$

$$C_w = W = Y - P \tag{22}$$

$$C_r = (1 - s_r)\{(1 - s_f)P + \rho V + i(\rho, d)D\} \tag{23}$$

[11] It should be noted that the model presented in this section, which is based on Asada (1999), is oversimplified and is not truly dynamic. However, Asada and Semmler (1995) presented a fully fledged, truly dynamic model of investment with a debt effect by means of Pontryagin's maximum principle, and derived a similar conclusion. The merit of the simplified model in this section may be its simplicity and transparency to clarify the basic idea.

[12] In this formulation, government economic activity and international trade are ignored for simplicity.

[13] However, we shall discuss some effects of price changes in Sect. 4.

where C_w is the workers' consumption, C_r is the capitalists' consumption, W is the wage income, P is the profit income, V is the value of the safe asset, and s_r is the capitalists' average propensity to save, which is assumed to be constant ($0 < s_r \leq 1$).

In this formulation, we consider the Kaleckian two-class model. Equation 22 implies that workers do not save, and that their consumption expenditure is exactly equal to their wage income. Equation 23 is the capitalists' consumption function. It is assumed that capitalists save a part of their income, and capitalists' income consists of three parts, i. e., the dividend ($(1 - s_f)P$), the interest receipt on the safe asset (ρV), and the interest receipt on the corporate debt ($i(\rho, d)D$).[14] Substituting Eqs. 6, 14, 21, 22, and 23 into Eq. 19, we obtain

$$\dot{y} = \alpha[\phi_1(g(r, \rho, s_f, d)) + (1 - s_r)\{\rho(V/K) + i(\rho, d)d\} - \{s_f + (1 - s_r)s_f\}r] \tag{24}$$

For income distribution, we simply assume that

$$P = \beta Y; \qquad 0 < \beta = \text{constant} < 1 \tag{25}$$

which implies that the "degree of monopoly," in the sense given by Kalecki (1971), is constant so that the share of profit in the national income is constant. In this case, we have the following expression of the rate of profit:

$$r = P/K = \beta Y/K = \beta y \tag{26}$$

Furthermore, for simplicity we assume that

$$V = \gamma D; \qquad \gamma = \text{constant} > 0 \tag{27}$$

Substituting Eqs. 26 and 27 into Eqs. 18 and 24, we obtain the following (nonlinear) system of fundamental dynamical equations with respect to the debt–capital ratio and the income–capital ratio.[15]

$$\begin{aligned} &\text{(i)} \quad \dot{d} = \phi_1(g(\beta y, \rho, s_f, d)) - s_f\{\beta y - i(\rho, d)d\} - g(\beta y, \rho, s_f, d)d \equiv f_1(d, y) \\ &\text{(ii)} \quad \dot{y} = \alpha[\phi_1(g(\beta y, \rho, s_f, d) + (1 - s_r)\{\rho\gamma + i(\rho, d)\}d - \{s_f + (1 - s_f)s_r\}\beta y] \\ &\qquad \equiv f_2(d, y) \end{aligned} \tag{28}$$

We now investigate the dynamic behavior of the system. We *assume* that this system has an equilibrium solution $(d^*, y^*) > (0,0)$. The Jacobian matrix of this system, which is evaluated *at the equilibrium point*, becomes

$$J \equiv \begin{bmatrix} f_{11} & f_{12} \\ f_{21} & f_{22} \end{bmatrix} \tag{29}$$

where

$$f_{11} \equiv \partial f_1/\partial d = \{\underset{(+)}{\phi_1'(g)} - d\}\underset{(-)}{g_d} - g + s_f\{\underset{(+)}{i_d}d + i\}$$

$$f_{12} \equiv \partial f_1/\partial y = \beta\{(\underset{(+)}{\phi_1'(g)} - d)\underset{(+)}{g_r} - s_f\}$$

[14] We assume that firms are debtors and capitalists are creditors.

[15] For simplicity, we assume that ρ is constant through time. This means that we are implicitly assuming that the monetary authority accommodates the money supply to the money demand passively to keep ρ constant.

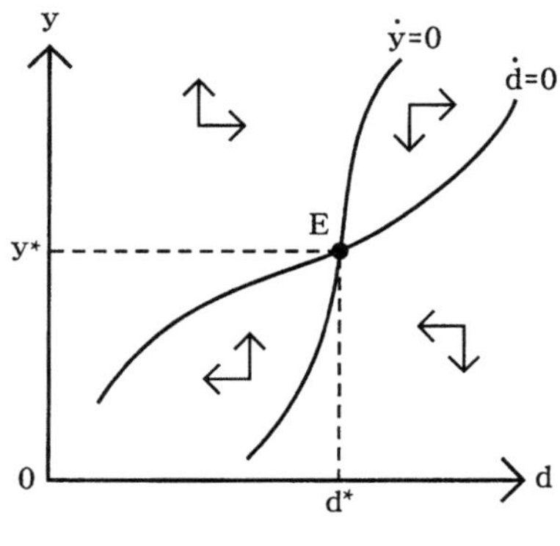

FIG. 2. Phase diagram

$$f_{21} \equiv \partial f_2/\partial d = \alpha[\underset{(+)}{\phi_1'(g)}\underset{(-)}{g_d} + (1-s_r)\{\underset{(+)}{i_d} d + \rho\gamma + i\}]$$

$$f_{22} \equiv \partial f_2/\partial y = \alpha\beta[\underset{(+)}{\phi_1'(g)}\underset{(+)}{g_r} - \{s_f + (1-s_f)s_r\}]^{16}$$

We now make the following assumptions.

Assumption 1: $f_{11} < 0, \quad f_{12} > 0, \quad f_{21} < 0, \quad f_{22} > 0$

Assumption 2: $f_{11}f_{22} - f_{12}f_{21} > 0$

The inequalities in Assumption 1 will be satisfied if $\phi_1'(g)$, g_r, and $|g_d|$ are sufficiently large at the equilibrium point. We can interpret the economic meaning of Assumption 2 as follows. We have

$$\lim_{s_r \to 1}(f_{11}f_{22} - f_{12}f_{21}) = \alpha\beta[\{-d\underset{(-)}{g_d} - g + (d\underset{(+)}{g_r} + s_f)(\underset{(+)}{i_d} d + i)\}\ \{\underset{(+)}{\phi_1'(g)}\underset{(+)}{g_r} - 1\} + s_f\underset{(+)}{\phi_1'(g)}\underset{(+)}{g_r}] \quad (30)$$

which will be positive if $\phi_1'(g)$, g_r, and $|g_d|$ are sufficiently large. This means that Assumption 2 will in fact be satisfied if $\phi_1'(g)$, g_r, and $|g_d|$, and s_r are sufficiently large at the equilibrium point.

Under these two assumptions, we have the following inequalities at the vicinity of the equilibrium point.

$$y'(d)|_{\dot{d}=0} = -(f_{11}/f_{12}) > 0, \qquad y'(d)|_{\dot{y}=0} = -(f_{21}/f_{22}) > 0$$
$$y'(d)|_{\dot{y}=0} - y'(d)|_{\dot{d}=0} = (f_{11}f_{22} - f_{12}f_{21})/f_{12}f_{22} > 0 \quad (31)$$

In this case, the local phase diagram in the vicinity of the equilibrium point is as shown in Fig. 2.

The following two propositions characterize the dynamics of the model in this section.

---

[16] We use the abbreviations $\partial g/\partial r = g_r$, $\partial g/\partial d = g_d$, and $\partial i/\partial d = i_d$.

FIG. 3. Closed orbit

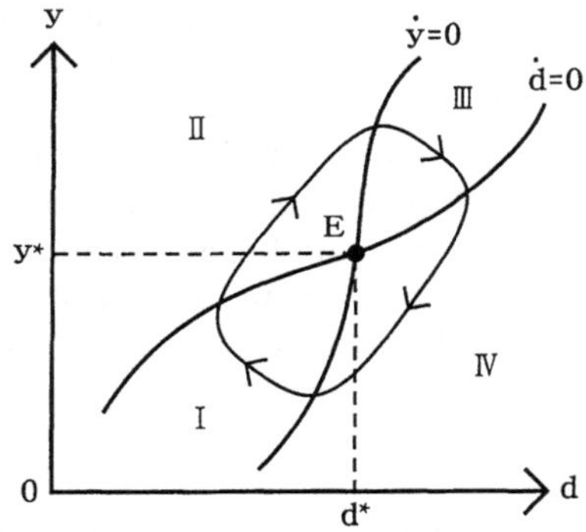

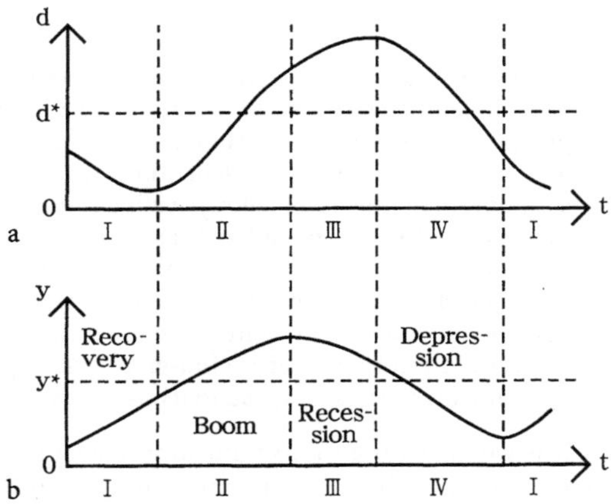

FIG. 4. Pattern of business cycles

Proposition 1: The equilibrium point of Eq. 28 is locally stable when $0 < \alpha < \alpha_0$, and it is locally *totally* unstable when $\alpha > \alpha_0$, where $\alpha_0 \equiv -f_{11}/\beta B > 0$ and $B \equiv \phi_1'(g)g_r - \{s_f + (1 - s_f)s_r\} > 0$.

See Appendix A for the proof of Proposition 1.

Proposition 2: At the parameter value $\alpha = \alpha_0$, the Hopf bifurcation occurs. In other words, there exist some nonconstant periodic solutions of the system Eq. 28 at some parameter values $\alpha > 0$ which are sufficiently close to α_0.

See Appendix B for the proof of Proposition 2.

Figure 3 illustrates an example of the closed orbit, the existence of which is ensured by Proposition 2. Figure 4 is the time trajectory of the solution corresponding to the four phases in Fig. 3.

In our model, the rate of change of the debt–capital ratio (d) is an increasing function of the income–capital ratio (y), and the rate of change of y is a decreasing function of d. This implies that the formal structure of our model is similar to the Volterra–Lotka system in mathematical biology, which describes the struggle between predator and prey.[17] In our model, d plays the role of predator and y plays the role of prey, and we can describe the scenario of the dynamics as follows. An increase in y induces debt-finance, and so induces an increase in d. However, an increase in the burden of debt induces a decrease in y through the decrease in investment expenditure, which will contribute to a decrease in d. If d decreases sufficiently, y will begin to increase again because of the increase in investment. In our model, the rate of capital accumulation (the rate of economic growth) fluctuates endogenously around the equilibrium rate g^*, while the equilibrium growth rate is also determined endogenously by the formula

$$g^* = g(\beta y^*, \rho, s_f, d^*) \tag{32}$$

Therefore, our model is an endogenous model of cyclical growth, and we can consider that it is a typical example of the formalization of the so-called "Minsky cycle" which was originated by Minsky (1975, 1986).[18] The basic idea is clearly described by Minsky (1975) as follows:[19]

> Financing is often based upon an assumption 'that the existing state of affairs will continue indefinitely' (GT, p. 152), but of course this assumption proves false. During a boom the existing state is the boom with its accompanying capital gain and asset revaluations. During both a debt-deflation and a stagnant recession the same conventional assumption of the present always ruling is made; the guiding wisdom is that debts are to be avoided, for debts lead to disaster. As a recovery approaches full employment the current generation of economic soothsayers will proclaim that the business cycle has been banished from the land and a new era of permanent prosperity has been inaugurated. . . . But in truth neither the boom, nor the debt-deflation, nor the stagnation, and certainly not a recovery of full-employment growth can continue indefinitely. Each state nurtures forces that lead to its own destruction. (p. 128)

It is worth noting that Proposition 2 ensures the *existence* of the closed orbit, but it says nothing about the *dynamic stability* of the closed orbit. It is well known that the stability of the closed orbit depends on the third-order partial derivatives of the relevant functions, but the economic interpretation of the stability condition is not clear. Therefore, in general, two cases are possible.

Figure 5a illustrates the case in which the bifurcation is "subcritical." In this case, the closed orbits exist at $\alpha < \alpha_0$, and the closed orbit is unstable although the equilibrium point is locally stable. This case corresponds to the "corridor stability" in the sense given by Leijonhufvud (1973). If the initial condition $(d(0), y(0))$ is inside the closed orbit (in the "corridor," so to speak), the solution converges to the equilibrium point, but if the initial condition is outside the "corridor," the system becomes dynamically unstable. In this case, the system is immune from "small" shocks, but it is

---

[17] An early example of the use of the Volterra–Lotka system in economic analysis is Goodwin (1967). See also Asada (1997), Keen (2000), and Lorenz (1993).

[18] The basic idea of this model is somewhat similar to the post-Keynesian models of financial dynamics which are referred to in footnote 3 in this paper, but our model is particularly close to the model by Delli Gatti and Gallegati (1995) in spirit, among others. In our model, which is formulated in terms of a two-dimensional continuous-time dynamic system, chaotic movement will not occur. However, Delli Gatti and Gallegati (1995) showed the possibility of chaotic movement by using the two-dimensional discrete-time model.

[19] The abbreviation "GT" in this quotation means Keynes' (1936) "general theory."

FIG. 5. Two types of bifurcation

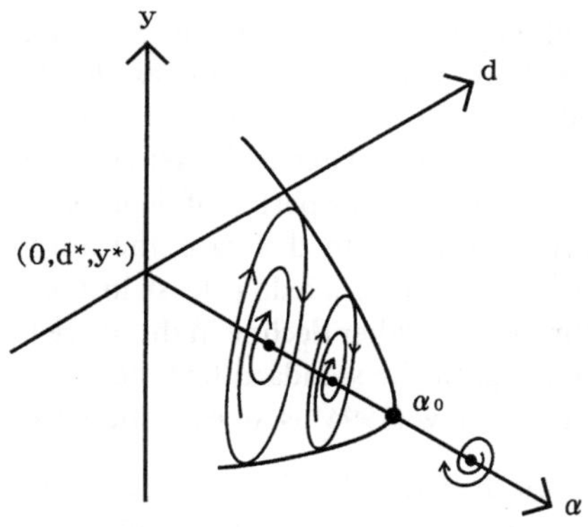

a

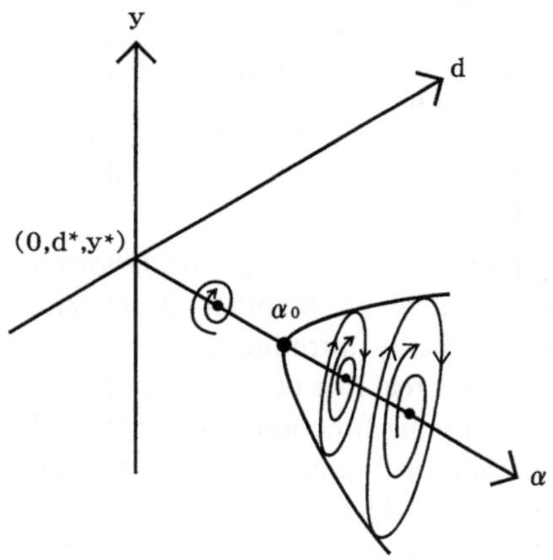

b

vulnerable to "large" shocks. On the other hand, Fig. 5b illustrates the case in which the bifurcation is "supercritical." In this case, the closed orbits exist at $\alpha > \alpha_0$, and the closed orbit is stable although the equilibrium point is locally unstable. In other words, the solution converges to the closed orbit if the initial condition is not at the equilibrium point, and the cyclical fluctuation will persist.[20]

Before closing this section, we briefly consider the *structural stability* of the system. Contrary to the original Volterra–Lotka system, which is structurally unstable, our system is structurally stable *except* at the point $\alpha = \alpha_0$. In other words, small changes in the parameter values will not change the qualitative nature of the system if $\alpha \neq \alpha_0$.

4 Concluding Remarks: Price Flexibility and Instability

So far, we have abstracted from the changes in prices. Now we are in a position to consider briefly the effect of price changes. The wisdom of the "classical" and "new classical" macroeconomics teaches us that price flexibility is enough to establish stable

[20] For subcritical and supercritical bifurcations, see Asada (1997) and Lorenz (1993).

full employment equilibrium. In fact, however, this is not necessarily the case. Price flexibility will contribute to *destabilize* rather than stabilize the economy if the "debt effect" is sufficiently strong. This is the main message of the "debt-deflation theory," which was originated by Fisher (1933) and reinterpreted by Keynes (1936) and Minsky (1975, 1986).[21] In a period of depression, the price levels will tend to decrease in a flexprice economy. In this case, the burden of the real debt will increase because the nominal debt is fixed, at least in the short run. This will aggravate the depression through a further decline in the aggregate demand. Without doubt, this effect is destabilizing. In the standard textbooks of macroeconomics, this fact is usually ignored, but Mankiw (1994) is an exception. He writes as follows:

> An unexpected deflation enriches creditors and impoverishes debtors. The debt–deflation theory then posits that this redistribution of wealth affects spending on goods and services. In response to the redistribution from debtors to creditors, debtors spend less and creditors spend more. If these two groups have equal spending propensities, there is no aggregate impact. But it seems reasonable to assume that debtors have higher propensities to spend than creditors—perhaps that is why the debtors are in debt in the first place. In this case, debtors reduce their spending by more than creditors raise theirs. The net effect is a reduction in spending, a contractionary shift in the IS curve, and lower national income. (Mankiw 1994 p. 279)

If we introduce price expectations into the model, another root of instability, which is known as the "Mundell effect," arises (cf. Flaschel et al. 1997). If price expectations are explicitly introduced, ρ in the previous section must be reinterpreted as the expected real rate of interest rather than the nominal rate of interest, and ρ becomes a decreasing function of the expected rate of price inflation (π^e). Even if the monetary authority keeps the *nominal* rate of interest constant, under the process of depression the reduction in π^e will raise ρ, which will induce the further decline of the effective demand through a decrease in the investment expenditure. Needless to say, this is also a destabilizing effect. Therefore, if we introduce price flexibility and price expectations into the model in the previous section, the economic fluctuations will be amplified rather than mitigated.[22]

Acknowledgment. This research was financially supported by Grant-in-Aid for Scientific Research No. 11630020 of the Japan Society for the Promotion of Science.

Appendix A: Proof of Proposition 1

The characteristic equation of the system (Eq. 28) becomes

$$|\lambda I - J| = \lambda^2 - (\text{trace}\, J)\lambda + \det J = 0 \tag{A1}$$

[21] See also Fazzari and Caskey (1989) and Tobin (1994).

[22] For formal analyses of instability in the flexprice economy with a debt effect, see Asada (2000) and Keen (2000).

where $\operatorname{trace} J = f_{11} + f_{22} = f_{11} + \alpha\beta B$ and $\det J = f_{11}f_{22} - f_{12}f_{21} > 0$. Then, the characteristic roots are given by

$$\lambda_1, \lambda_2 = \left\{\operatorname{trace} J \pm \sqrt{(\operatorname{trace} J)^2 - 4\det J}\right\}/2 \tag{A2}$$

Suppose that $0 < \alpha < \alpha_0$. In this case, we have $\operatorname{trace} J < 0$, which implies that the real parts of both roots become negative irrespective of the sign of $(\operatorname{trace} J)^2 - 4\det J$, so that the equilibrium point of Eq. 28 becomes locally stable.

Next, suppose that $\alpha > \alpha_0$. In this case, we have $\operatorname{trace} J > 0$, which implies that the real parts of both roots become positive, so that the equilibrium point of Eq. 28 becomes locally totally unstable. □

Appendix B: Proof of Proposition 2

Suppose that $\alpha = \alpha_0$. In this case, we have $\operatorname{trace} J = 0$, so that the characteristic roots become

$$\lambda_1, \lambda_2 = \pm i\sqrt{\det J} \tag{B1}$$

where $i = \sqrt{-1}$, which implies that the characteristic equation has a pair of pure imaginary roots at $\alpha = \alpha_0$. We also have

$$d\{\operatorname{Re}\lambda(\alpha_0)\}/d\alpha = d(\operatorname{trace} J/2)/d\alpha = \beta B/2 > 0 \tag{B2}$$

where $\operatorname{Re}\lambda(\alpha_0)$ is the real part of the characteristic roots when $\alpha = \alpha_0$. Furthermore, the equilibrium values (d^*, y^*) are independent of the value of the parameter α. This proves that the point $\alpha = \alpha_0$ is in fact the Hopf bifurcation point (see Appendix C). □

Appendix C: Hopf Bifurcation Theorem

The proof of Proposition 2 is based on the following version of the Hopf bifurcation theorem (cf. Asada 1997; Asada and Semmler 1995; Flaschel et al. 1997; Lorenz 1993).

Theorem

Let $\dot{x} = f(x; \varepsilon)$, $x \in R^n$, $\varepsilon \in R$ be a system of differential equations with a parameter ε. Suppose that the following properties (i)–(iii) are satisfied. (i) This system has a smooth equilibria curve $f(x^*(\varepsilon); \varepsilon) = 0$. (ii) The characteristic equation $|\lambda I - Df(x^*(\varepsilon_0); \varepsilon_0)| = 0$ has a pair of pure imaginary roots $\lambda(\varepsilon_0)$, $\bar{\lambda}(\varepsilon_0)$ and no other roots with zero real parts, where $Df(x^*(\varepsilon_0); \varepsilon_0)$ is the Jacobian matrix of the above system at $(x^*(\varepsilon_0), \varepsilon_0)$ with the parameter ε_0. (iii) $d\{\operatorname{Re}\lambda(\varepsilon)/d\varepsilon|_{\varepsilon=\varepsilon_0} \neq 0$, where $\operatorname{Re}\lambda(\varepsilon)$ is the real part of $\lambda(\varepsilon)$.

Then, there exists a continuous function $\varepsilon(\gamma)$ with $\varepsilon(0) = \varepsilon_0$, and for all sufficiently small values of $\gamma \neq 0$ there exists a continuous family of nonconstant periodic solutions $x(t, \gamma)$ for the above dynamic system, which collapses to the equilibrium point $x^*(\varepsilon_0)$ as $\gamma \to 0$. The period of the cycle is close to $2\pi/\operatorname{Im}\lambda(\varepsilon_0)$, where $\operatorname{Im}\lambda(\varepsilon_0)$ is the imaginary part of $\lambda(\varepsilon_0)$.

References

Arena R, Raybaut A (1998) Credit and financial markets in Keynes' conception of endogenous business cycles: an interpretation. In: Abraham-Frois G (ed) Non-linear dynamics and endogenous business cycles. Springer, Berlin, Heidelberg, New York, pp 147–116

Asada T (1997) Macrodynamics of growth and cycles (in Japanese). Nihon Keizai Hyoronsha, Tokyo

Asada T (1999) Investment and finance: a theoretical approach. Ann Oper Res 89:75–87

Asada T (2000) Debt effect and macroeconomic instability. Annals of the Institute of Economic Research, Chuo University (Tokyo), vol 31, forthcoming

Asada T, Semmler W (1995) Growth and finance: an intertemporal model. J Macroecon 17:623–649

Baumol WJ, Heim P, Malkiel BJ, et al. (1970) Earnings retention, new capital and growth of the firm. Rev Econ Stat 52:345–355

Bernanke B, Gertler M (1989) Agency costs, net worth, and business fluctuations. Am Econ Rev 79:14–31

Delli Gatti D, Gallegati M (1994) External finance, investment expenditure and the business cycle. In: Semmler W (ed) Business cycles: theory and empirical methods. Kluwer, Boston, Dordrecht, London, pp 269–288

Delli Gatti D, Gallegati M (1995) Financial fragility and economic fluctuations: Keynesian views (and an addendum). Econ Notes 24:513–554

Fazzari S, Caskey J (1989) Debt commitments and aggregate demand: a critique of the neoclassical synthesis and policy. In: Semmler W (ed) Financial dynamics and business cycles: new perspectives. Sharpe, New York, pp 188–199

Fazzari S, Hubbard G, Petersen BC (1988) Financing constraint and corporate investment. Brookings Pap Econ Act 1:141–206

Fisher I (1933) The debt–deflation theory of great depressions. Econometrica 1:337–357

Flaschel P, Semmler W, Franke R (1997) Dynamic macroeconomics: instability, fluctuations, and growth in monetary economies. MIT Press, Cambridge

Franke R, Semmler W (1989) Debt-financing of firms, instability, and cycles in a dynamical macroeconomic growth model. In: Semmler W (ed) Financial dynamics and business cycles: new perspectives. Sharpe, New York, pp 38–64

Goodwin RM (1967) A growth cycle. In: Feinstein CH (ed) Socialism, capitalism and economic growth. Cambridge University Press, Cambridge, pp 54–58

Greenwald BC, Stiglitz JE (1993) Financial market imperfections and business cycles. Q J Econ 108:77–114

Jarsulic M (1990) Debt and macrostability. East Econ J 15:91–100

Kalecki M (1937) The principle of increasing risk. Economica 4:440–447

Kalecki M (1971) Selected essays on the dynamics of the capitalist economy. Cambridge University Press, Cambridge

Keen S (2000) The nonlinear economics of debt deflation. In: Barnett WA, Chiarella C, Keen S, Marks R, Schnabl H (eds) Commerce, complexity, and evolution. Cambridge University Press, Cambridge, pp 83–110

Keynes JM (1936) The general theory of employment, interest and money. Macmillan, London

Kuroki R (1994) Growth and debt accumulation. In: Aoki N, Shiraiwa K, Takahashi Y (eds) Proceedings of the International Conference on Dynamical Systems and Chaos. World Scientific, Singapore, New Jersey, London, Hong Kong, pp 399–403

Leijonhufvud A (1973) Effective demand failure. Swed J Econ 75:27–48

Lorenz HW (1993) Nonlinear dynamical economics and chaotic motion. Springer, Berlin, Heidelberg, New York

Mankiw NG (1994) Macroeconomics, 2nd edn. Worth, New York

Minsky HP (1975) John Maynard Keynes. Columbia University Press, New York

Minsky HP (1986) Stabilizing an unstable economy. Yale University Press, New Haven
Modigliani F, Miller M (1958) The cost of capital, corporation finance and the theory of investment. Am Econ Rev 53:261–297
Pally T (1996) Post-Keynesian economics: debt, distribution and the macroeconomy. Macmillan, London
Steigum E (1983) A financial theory of investment behavior. Econometrica 51:637–645
Steindl J (1945) Capitalist enterprise and risk. Oxford Economic Papers No. 7. Reprinted In: Steindl J, Econ Pap 1941–1988. Macmillan, London, pp 13–36
Taylor L, O'Connell SA (1985) A Minsky crisis. Q J Econ 100(Suppl):871–886
Tobin J (1994) Price flexibility and output stability: an old Keynesian view. In: Semmler W (ed) Business cycles: theory and empirical methods. Kluwer, Boston, Dordrecht, London, pp 165–195
Uzawa H (1969) Time preference and the Penrose effect in a two-class model of economic growth. J Polit Econ 628–652

Attractor Stability in Unemployment and Inflation Rates*

Stephen J. Guastello

Summary. Several discrepancies have surfaced recently regarding the structure of unemployment and inflation rates. These discrepancies can be translated into dubious national policies if they are left unchecked. The research reported here defines the controversial points as hypotheses in nonlinear dynamics. Statistical tests for chaos, fixed-point attractors, and bifurcation structures indicated the presence of two asymptotic points in unemployment rates in the US, one of which was a chaotic attractor and the other an unstable point resulting from a bifurcation structure. Similar tests indicated the presence of chaotic attractors for inflation rates with nonzero epicenters.

Key words. Unemployment, Inflation, Chaos, Attractor, Bifurcation

1 Problem Statement

After Keynes' (1936/1965) examination of the relationships that exist between rates of unemployment and inflation, a nonlinear relationship between the two quantities was articulated in the 1950s in the form of the Phillips (1954) curve. By the late 1960s, however, the curve was found not to hold up against data that had been accumulated (Phelps 1967), and thus it was replaced by the concept of a natural rate of unemployment (Friedman 1968). This so-called natural rate was conceptualized as a Walrasian equilibrium, which is known to dynamicists as a fixed-point attractor (Dore 1998). Unfortunately, economic policy in the US in the 1990s continued to regard unemployment as a voluntary market function, and to define the natural rate of unemployment as whatever the current rate happened to be (Freedman and Kriesler 1994; Seccareccia 1991). Thus, in practice, the natural rate theory was a universal justification for any abnormally high unemployment rates, which had reached a local peak of between 10% and 11% during the 1989–1995 period.

Department of Psychology, Marquette University, PO Box 1881, Milwaukee, WI 53201-1881, USA

* Portions of this text and figures appeared in Guastello (1995b) and Guastello (1999), and were reprinted by permission. Please see reference list for citations.

Lerner (1951), on the other hand, suggested that the actual level of full employment should be targeted relative to the inflation rate. If inflation is already low, then higher full employment is possible without triggering high rates of inflation. Income policy was thought to be the key variable for controlling both inflation and unemployment. To Sraffa (1960), however, unemployment levels were far less connected to market forces than they were to power relationships between employer and employee. In a no-growth situation, an employer would seek to increase profits by pressuring wages. Wages could be pressured directly by lowering wages, or indirectly by forcing a higher turnover and bidding down the next wave of job applicants.

A natural rate of unemployment, should it exist, is a convenient assumption for national-level policy makers who adjust central bank interest rates on the basis of money supply, inflation rates, unemployment, and international currency exchange trends. Yet observers of US federal monetary policy have expressed concern that the central bank has had no consistent policy in its adjustments to interest rates during the 1985–1995 decade (Galbraith 1994; Papadimitriou and Wray 1994). From one vantage point, the focus of policy may have shifted from monitoring price stability to managing employment levels (Davidson 1995). In other words, there seemed to be enough residual belief in the Phillips curve for policy makers to assume that if unemployment becomes too low, inflation must be "right around the corner." It did appear, furthermore, that hikes in interest rates were deliberate attempts to generate more unemployment.

Policy makers in the US were much less quick to assert the existence of a natural rate of inflation. Instead, policies appeared to be directed toward zero inflation, which according to Keynes (1965) was technically impossible. The idea of a natural rate of inflation was actually proposed by two Australian economists (Freedman and Kriesler 1994).

Other theorists suggested different dynamics for unemployment and inflation. According to Schultze (1959), increased wages are often seen as the culprit of inflation, but during some epochs, increased wage demands are responses to increased corporate profits. Mitchell (1993) found that unemployment rates were generally stationary for most Organization for Economic Cooperation and Development (OECD) countries, but they displayed hysteresis in others. Sraffa, on the other hand, characterized an economy as an intense network of purchases and sales; in contemporary thinking, any cyclical behavior among the economic sectors could result in chaotic dynamics or dampening of the cycles altogether (e.g., Puu 1993).

The research reported here examined the various interpretations of unemployment and inflation dynamics as hypotheses in nonlinear dynamic systems theory. This report necessarily assumes some familiarity with fixed-point attractors, periodic attractors, chaos, and bifurcation. The interested reader is referred to Puu (2000) and Rosser (1991, 2000) for explanations of how those dynamics play out in a variety of economic phenomena.

The analyses presented here utilized the method of exponential nonlinear regression equations (Guastello 1995a). With this procedure, it was possible to detect fixed points, chaos, and bifurcation effects. US data for 1948 until early 1995 were used. The hypotheses are given below.

The natural rate explanation suggests that unemployment rates have a fixed-point attractor.

Policies based on a natural-rate assumption or a loose Keynesian mechanism suggest that instabilities in unemployment are the norm, although some instabilities are greater than others.

The complex system of economic forces that affect unemployment rates can produce both upward and downward effects. This pushing and pulling suggests chaotic dynamics.

Policies that target zero inflation assume that zero inflation is an optimal equilibrium value.

According to Keynesian theory and the horizontal Phillips model (Freedman and Kriesler 1994), a nonzero equilibrium must exist for inflation.

Instabilities in inflation rates are known to have occurred in some countries; they could be evident in the US as well.

What are the critical values of employment and inflation stability and instability if they exist?

2 Analytical Method

The primary tool for analyzing dynamics is the Lyapunov dimension (D_L). D_L is an indicator of the amount of exponential expansion that is taking place over time (Kaplan and Glass 1995); it provides both a test of chaos and an indicator of system complexity. Nonstatistical models can be restructured as a calculation of a from a time series of a dependent variable (order parameter) y such that

$$y = \|e^{at}\| \tag{1}$$

(Guastello 1995a; Wiggins 1988). If a is positive, then the function is chaotic, expanding, and sensitive to initial conditions. If a is negative, then the function is a fixed point or periodic attractor, contracting, and insensitive to initial conditions. D_L is then calculated as

$$D_L = e^a \tag{2}$$

The most workable statistical method for calculating D_L utilizes nonlinear regression coupled with structural modeling (Guastello 1995a). The process begins by modeling the simplest structure:

$$z_2 = e^{\theta_1 z_1 t} + \theta_2 \tag{3}$$

where θ_1 is the critical exponent, θ_2 is a constant (of little consequence), and z_2 and z_1 are consecutive values of our dependent measure in a time series that has been corrected for location (λ) and scale (σ_s).

$$z = (y - \lambda)/\sigma_s \tag{4}$$

(Guastello 1992, 1995a). The resulting time series variable z is now calibrated in moments rather than in its original units. Also t = time in Eq. 3, and because all time lapses are equal intervals, t is set to the trivial value of 1. The resulting measure of dimension is then, following Eq. 2, $D_L = e^{\theta_1}$.

The second model in the series is the case where we have an unknown bifurcation variable. Here we test for the structure, but since we have neither a hypothesis as to what it could be or data to test it, we estimate the bifurcation variable θ_1 as a regression parameter:

$$z_2 = \theta_1 z_1 e^{\theta_2 z_1} + \theta_3 \tag{5}$$

In this case, dimension is calculated as

$$D_L = e^{\theta_2} + 1 \tag{6}$$

The next step in the assessment of nonlinear trends in order to compare the nonlinear functions with linear alternative explanations. In other circumstances, where the data contain N objects measured at two points in time, the following linear model is a good comparison:

$$z_2 = b_1 z_1 + b_2 x_1 + b_2 x_2 + b_n x_n \tag{7}$$

In Eq. 7, z is the dependent measure at two points in time as before, b_i are ordinary regression weights, and x_i are other independent variables which we do not have in this application. The task then is to compare R^2 values for the linear and nonlinear models.

The sampling frequency will affect the size of the R^2 of the linear model relative to the R^2 of the nonlinear model. For short lag lengths, the trajectory between two points in time is well approximated by a linear model. It is only when the global picture is taken into account that we see the nonlinear structure that could be taking shape. Equation 7, when applied to locally smooth trends, can produce an accurate forecasting model (or retrocasting model) that says that z_2 is a linear function of z_1; fresh observations of z_1 need to be made (almost) every time the forecast is to be moved ahead by one time interval.

Another alternative comparison model is to regress z against time:

$$z_1 = bt + C \tag{8}$$

Equation 8 says that z is a linear function of time and is not further dependent on its previous value. This is probably a fairer comparison from a modeling standpoint, but Eq. 7 may be preferable from the viewpoint of prediction.

3 Analysis for Unemployment

The annualized unemployment trends for the 1948–1994 period are shown in Fig. 1. The results for the simplest model were

$$z_2 = e^{(0.36 z_1)} - 0.205 \tag{9}$$

for which values of $R^2 = 0.86$ and Lyapunov dimension (D_L) = 1.43 were obtained. This is interpreted as showing that the function is low-dimensional chaos, as denoted by the positive exponent. This chaotic effect may reflect an autonomous process for which no control parameter is really responsible. Exogenous forces, e.g., an occasional political decision, may be responsible for specific random shocks which do not send trajectories of unemployment rates out of their attractor basin. However, such shocks are carried through the subsequent portion of the time series. The system is structurally stable.

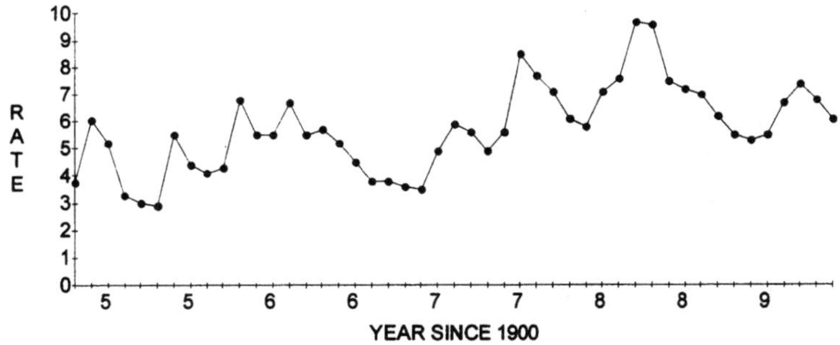

FIG. 1. Annualized unemployment rates for the US 1948–1994 (From Guastello 1999, with permission)

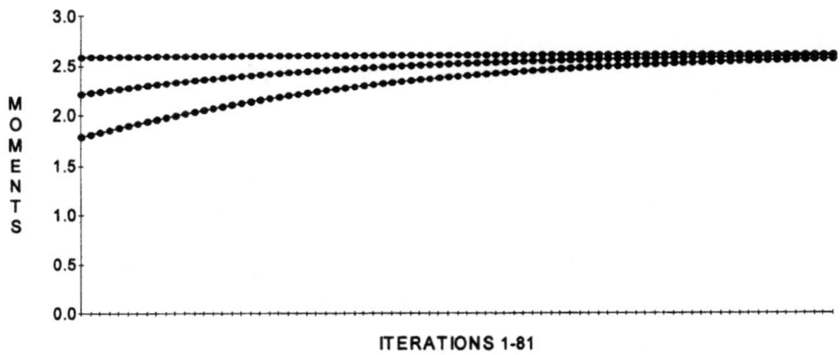

FIG. 2. Iterations of the bifurcation model based on the high, average, and low monthly unemployment rates for 1994 as initial conditions (From Guastello 1999, with permission)

The results for the bifurcation model were

$$z_2 = 1.04z_1e^{(-0.015z_1)} \tag{10}$$

The values of $R^2 = 0.92$ and $D_L = 1.98$ were obtained. There was a significant bifurcation factor, which meant that a factor is present that was responsible for the widening and contracting of the periodic fluctuations. Results thus favored the logistic bifurcation interpretation.

For the linear comparison, where z_2 is tested as a function of z_1, an R^2 of 0.92 was obtained. For the linear comparison where z_2 was tested as a function of time, an R^2 of 0.32 was obtained. Both nonlinear models were superior to linear functions of time.

Because the logistic had a negative exponent, iteration was initialized with the average unemployment rate for 1994, and then run until it was close to convergence (48 iterations, analogous to 12 years of real time). All the usual assumptions and risks concerning ergodicity and forecasting apply. The natural rate obtained from this situation was 6.6% of the workforce unemployed. The iterated time series appears in Fig. 2.

Because the chaotic attractor function had a positive exponent, the method for finding the epicenter of the attractor is to multiply the exponent by $-1t$, and iterate

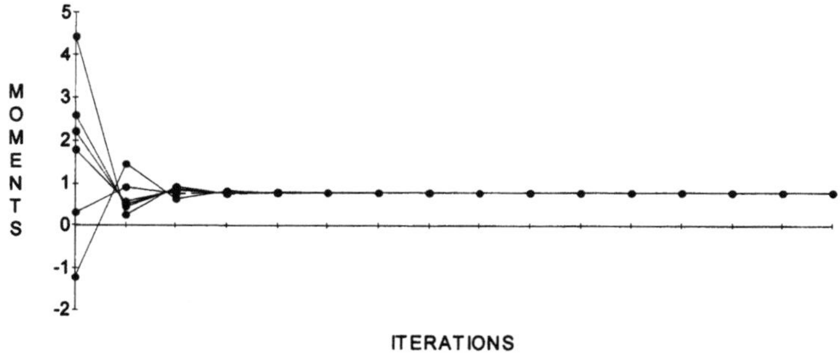

FIG. 3. Backward iteration of the chaotic attractor function to locate its epicenter (From Guastello 1999, with permission)

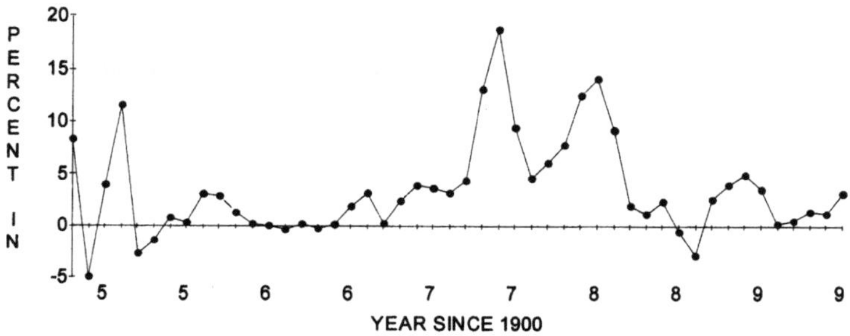

FIG. 4. Annualized inflation rates for the PPI, US, 1948–1995 (From Guastello 1995b, with permission)

backwards in time to find the initial conditions that are most likely to have produced the time series we observe (Wiggins 1988). That value was 3.4% unemployment. The iterated time series for the chaotic attractor function appears in Fig. 3.

4 Analysis for Inflation

4.1 Producer Price Index (PPI)

Annualized data for the PPI for the US from 1948 to 1995 appear in Fig. 4. A chaotic attractor was isolated for the PPI ($R^2 = 0.75$) with $D_L = 1.51$ at a 3-month lag interval. The resulting function was

$$z_2 = e^{(0.411z_1)} - 0.789 \tag{11}$$

The attempt to test a bifurcation model resulted in a nonsignificant exponent, such that the null hypothesis of linearity was upheld. However, the time series was well approximated by the linear autoregressive function, whereby fluctuations were dampened from quarter to quarter and a small increase in inflation was added ($R^2 = 0.84$). The test for the linear effect over time showed a poor level of fit, by contrast ($R^2 = 0.02$).

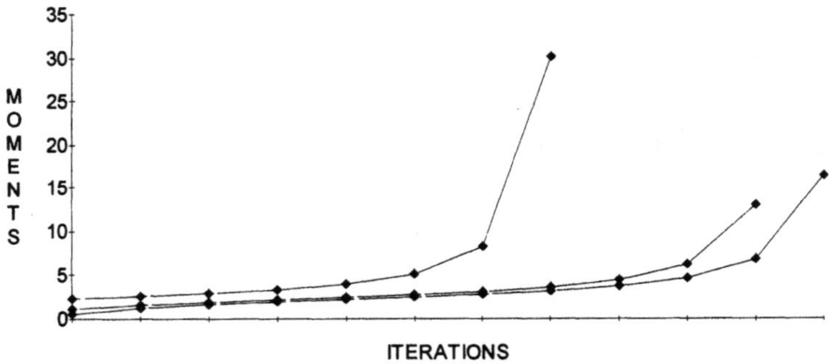

FIG. 5. Chaotic function for PPI iterated forward in time without constant (From Guastello 1995b, with permission)

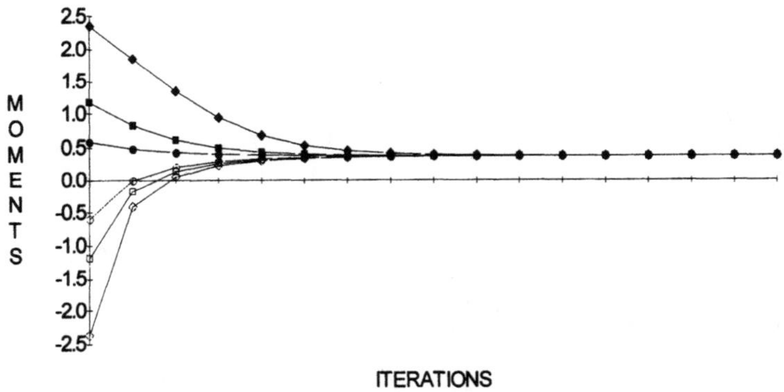

FIG. 6. Chaotic function for PPI iterated forward in time with damping constant (From Guastello 1995b, with permission)

The next procedure was to explore the function obtained for the PPI forecast outcomes based on plausible initial conditions, and reverse-forecast the function to determine plausible initial conditions that might have produced it. Because D_L is evaluated at the norm of the time series, the constant θ_2 can be interpreted as a further correction for location. Thus, iterations would proceed without the constant, after which the constant would be added to z_i before multiplying by the scale parameter to obtain real values, y_i.

Depending on the specifics of the problem under study, however, θ_2 may represent the norm of an opposing function. The inflation problem may be such an example inasmuch as the sources of price increases at the producer level would trigger feedback to the producer to engage in cost-cutting measures. Thus, iterative series were explored with and without the constant.

Figure 5 shows the forecasts for the PPI based on its chaotic statistical model without the constant, with initial conditions of 3%, 6%, and 12% inflation. As expected, inflation rates rise in a nearly linear fashion until they hit a critical value and then zoom off the graph into rampant inflation. Figure 6 shows forecasts for the PPI based on the same chaotic function with the constant included, the same three conditions of inflation, plus

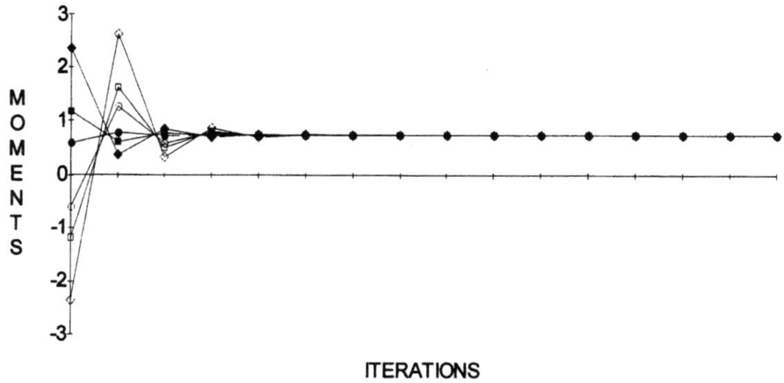

FIG. 7. Chaotic function for PPI iterated backward in time to determine its epicenter (From Guastello 1995b, with permission)

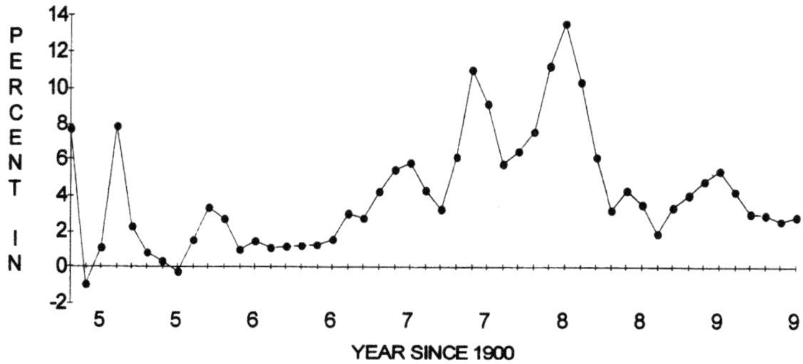

FIG. 8. Annualized inflation rates for the CPI, US, 1948–1995 (From Guastello 1995b, with permission)

initial conditions of 3%, 6%, and 12% deflation. In all cases, iterates reached asymptotic stability at 0.38 moments, which correspond to an inflation rate of 1.93%.

Next, a backward iteration was performed to identify a plausible initial condition that could have produced the attractor that was obtained. Here, all six starting conditions converged to an epicenter of 0.74 moments (Fig. 7). After subtracting the θ_2 constant, the epicenter corresponded to –0.26%. This value represents a near-balance of forces affecting price increases and decreases, with a slight tendency toward price decreases.

4.2 Consumer Price Index (CPI)

Annualized data for the CPI for the US from 1948 to 1995 appear in Fig. 8. A chaotic attractor was isolated for the CPI ($R^2 = 0.85$) with $D_L = 1.49$ at a 3-month lag interval (554 observations). The resulting function was

$$z_2 = e^{(0.398 z_1)} - 0.555 \tag{12}$$

The test of the bifurcation model produced an exponent that was not significantly different from zero; therefore, the null hypothesis of linearity was upheld for that model. However, the time series was well approximated by the linear autoregressive function, whereby fluctuations were dampened from quarter to quarter and a small increase in inflation was added ($R^2 = 0.90$). The test for the linear effect over time showed a poor level of fit, by contrast ($R^2 = 0.40$).

To explore the implications of the chaotic function, the function was iterated backwards without the constant to identify the epicenter of the attractor. Asymptotic stability was obtained at $z = 0.324$ moments, which, after subtracting the constant and multiplying by the scale parameter, resulted in an epicenter value of –0.76%. This value represented a small downward push on prices. The results are shown in Fig. 9 for initial conditions of 3%, 6%, and 12% inflation.

The chaotic function was then iterated forward in time with the constant, using the same set of initial conditions. For initial conditions of 3% and 6%, the function converged asymptotically to 0.84 moments, corresponding to 2.78% inflation. For initial conditions of 12% inflation, however, there was a surprise—inflation grew slowly, then took off at runaway proportions. Figure 10 shows a bit a surfing that was undertaken

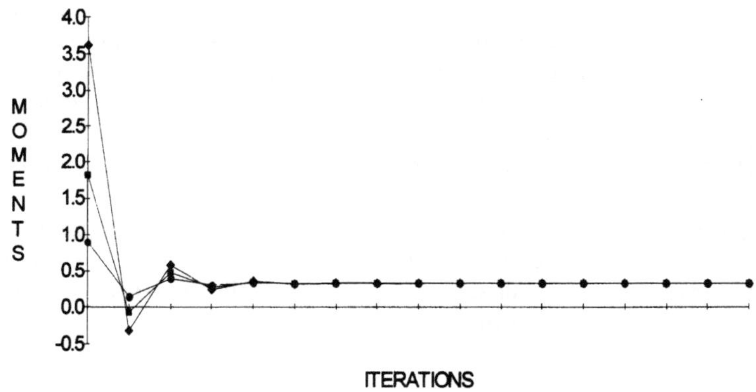

FIG. 9. Chaotic function for CPI iterated backward in time to determine its epicenter (From Guastello 1995b, with permission)

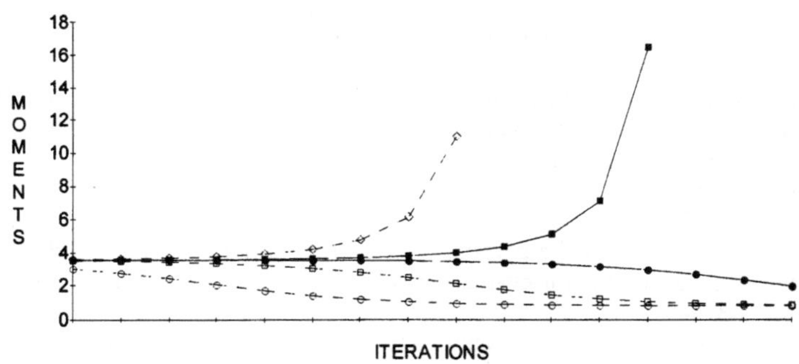

FIG. 10. Chaotic function for CPI iterated forward in time, showing the critical point (From Guastello 1995b, with permission)

to determine the critical initial condition that would be responsible for inflation containment and runaway conditions. The critical value was between 3.55 and 3.56 moments, corresponding to initial conditions of 11.75% and 11.78% inflation, respectively.

5 Discussion

The results of this study supported the following conclusions. Unemployment rates do not have a fixed-point attractor, and therefore are not "natural rates" in the sense of fixed-point attractors. The best characterization of unemployment rates identified a bifurcation effect, which denotes instabilities over time. Rates in this model were inclining through 6.6%, but were not expected to stabilize.

The second-best characterization of the unemployment data indicated a chaotic attractor with an epicenter at 3.4%. The chaotic structure may be interpreted as the closest alternative to a fixed-point natural rate, but of course, it is not fixed in any sense.

A nonzero equilibrium in price inflation exists; the structure is chaotic and may be interpreted as the closest alternative to a fixed-point natural rate. The epicenter of the PPI attractor is located at 0.26%, signifying a slight downward pressure on prices. If the PPI rate is iterated forward in time, it stabilizes at a rate of 1.93%. The forward iteration with the negatively valued constant appears to represent a boundary condition. The boundary of the attractor basin is probably held in place by a combination of cost-cutting initiatives from producers of goods and federal economic policy.

The epicenter of the CPI attractor is located at –0.76%, again signifying a slight downward pressure on prices. If the CPI rate is iterated forward in time with the negatively valued constant, it stabilizes at a rate of 2.78%. If the CPI reaches a critical point between 11.75% and 11.78%, runaway inflation will occur.

References

Davidson P (1995) Letter to the editor. New York Times. Electronically retrievable document by File Transfer Protocol: PKT Archives, January 25, 1995: csf.colorado.edu.

Dore MHI (1998) Walrasian general equilibrium and nonlinear dynamics. Nonlinear Dyn Psychol Life Sci 2:59–72

Freedman C, Kriesler P (1994) Has the long-run Phillips curve turned horizontal? Technical Report No. 94/27, School of Economics, University of New South Wales, Sydney

Friedman M (1968) The role of monetary policy. Am Econ Rev 58:1–17

Galbraith JK (1994) Self-fulfilling prophets: inflated zeal at the Federal Reserve. Am Prospect 18:31–39

Guastello SJ (1992) Clash of the paradigms: a critique of an examination of the polynomial regression technique for evaluating catastrophe theory hypotheses. Psychological Bulletin 111:375–379

Guastello SJ (1995a) Chaos, catastrophe, and human affairs: applications of nonlinear dynamics to work, organizations, and social evolution. Lawrence Erlbaum, Mahwah

Guastello SJ (1995b) The search for a natural rate of price inflation: US 1948–1995. Chaos Network 7(3):16–23

Guastello SJ (1999) Hysteresis, bifurcation, and the natural rate of unemployment. In: Elliott E, Kiel LD (eds) Nonlinear dynamics, complexity and public policy. Nova Science, Commack, pp. 31–46
Kaplan D, Glass L (1995) Understanding nonlinear dynamics. Springer, New York
Keynes JM (1965) General theory of employment, interest, and money, 2nd edn. Harcourt Brace, New York
Lerner A (1951) Economics of employment. McGraw-Hill, New York
Mitchell WF (1993) Testing for unit roots and persistence in OECD unemployment rates. Appl Econ 25:1489–1501
Papadimitriou DB, Wray LR (1994) Flying blind: The Federal Reserve's experiment with unobservables. Technical Report, Public Policy Brief, Jerome Levy Economic Institute, Bard College
Phelps ES (1967) Phillips curves, expectations of inflation and optimal unemployment. Economica 34:254–281
Phillips AW (1954) Stabilisation policy in a closed economy. Norton, London
Puu T (1993) Nonlinear economic dynamics, 3rd edn. Springer, New York
Puu T (2000) Attractors, bifurcations, and chaos: nonlinear phenomena in economics. Springer, New York
Rosser JB Jr (1991) From catastrophe to chaos: a general theory of economic discontinuities. Kluwer, Boston
Rosser JB Jr (2000) From catastrophe to chaos: a general theory of economic discontinuities, 2nd edn. Vol 1. Mathematics, microeconomics, and finance. Kluwer, Boston
Schultze CL (1959) Prices, costs, and output for the postwar decade: 1947–1957. Center for Economic Development, New York
Seccareccia M (1991) An alternative to labour-market orthodoxy: the post-Keynesian/institutionalist policy view. Rev Polit Econ 3:43–61
Sraffa P (1960) Production of commodities by means of commodities: prelude to a critique of economic theory. Cambridge University Press: London
Wiggins S (1988) Global bifurcations and chaos. Springer, New York

Part III Experimental Economics and Evolution

Evolution and Negative Reciprocity

Daniel Friedman* and Nirvikar Singh*

Summary. We offer a theoretical explanation of negative reciprocity or vengeance, i.e., the human desire to harm those who have harmed us. Our model shows how negative reciprocity can be sustained by the coevolution[1] of genes that determine the capacity for vengeance and group memes (e.g., social norms) that regulate its expression. The model begins with a standard free-rider game that captures, simply and directly, a personal cost incurred to reap social gains. The model shows that a taste for vengeance realigns incentives and supports a socially efficient equilibrium, but that by itself the taste for vengeance is not evolutionarily viable. We then show how groups of individuals can use low-power sanctions (or simply status changes) to enforce a particular norm on the proper degree of vengeance. The main result is that actual behavior will typically fall short of the norm, but selection across groups will adjust the norm so that actual behavior maximizes the fitness of group members.

Key words. Negative reciprocity, Vengeance, Coevolution, Memes

1 Introduction

Vengeance is a powerful human motive. We become angry when someone wrongs us, and often try to harm the culprit in return, even at some personal cost. Vengeance deters opportunistic behavior[2] that otherwise might undermine positive reciprocity, direct or indirect,[3] and thereby supports social cooperation. On the other hand, misplaced vengeance sometimes leads to tragic feuds and even genocide, as in Rwanda or the Balkans.[4]

---

* Department of Economics, Social Sciences 1, University of California at Santa Cruz, Santa Cruz, CA 95064, USA

[1] See Dawkins (1976), Boyd and Richerson (1985), and Durham (1991).
[2] See, for example, Fehr and Gaechter (1999).
[3] See, for example, Nowak and Sigmund (1998).
[4] See, for example, Hagen (1999).

The existence of vengeance is empirically obvious (and confirmed in controlled experiments),[5] but theoretically mysterious, because vengeance is not individually rational: it is weakly dominated by otherwise similar behavior that avoids the personal cost. Therefore it is a theoretical puzzle how vengeance ever established itself in the repertoire of human motives, and how it sustains itself. Until the puzzle is solved, theory will offer no guidance on how to regulate vengeance to maximize its social value and to minimize its devastation.

2 Vengeance Solves Social Dilemmas

We begin our analysis with the simplest possible social dilemma, written as a symmetric two-player game (Table 1). The entries in Table 1 denote fitness payoffs to a player choosing that row (C = cooperate; D = defect) when the partner chooses that column (again C or D). Thus the personal cost (fitness reduction) of choosing C rather than D is 1, and the social gain (own payoff plus partner's payoff) is also 1. The game has a unique equilibrium: each player chooses the dominant strategy D and achieves fitness 0, thus missing the potential gains of 1 for each player.

We now add a punishment technology and a punishment motive, parameterized by its incurred cost v. We hypothesize that a player can inflict harm (fitness loss) h on the other player at a personal fitness cost ch. The marginal cost c in (0,1) is a constant parameter that captures the technological opportunities for punishing others.

In addition, inflicting harm h yields the player a utility bonus of $v \ln h$ (but no fitness bonus) when he is the victim of the sucker payoff (receiving −1, while the culprit receives +2) and no bonus in other circumstances. Thus, the motive is not spite,[6] but rather is vengeance for damage personally experienced. The motivational parameter v is subject to evolutionary forces and captures an individual's temperament, e.g., his susceptibility to anger.[7]

The objective function for the victim of a sucker payoff now is $v \ln h - ch - 2$. Then $h^* = v/c$ is the utility-maximizing degree of inflicted damage, and $ch^* = v$ the incurred cost. The game is now as shown in Table 2. For $v > c$, the strategy D is not dominant. When population fraction s plays C, the expected fitness of C is $W(C) = 1s - (1 + v)(1 - s)$, and the expected fitness of D is $W(D) = (2 - v/c)s$. The two expressions

TABLE 1. Fitness with no vengeance

$(v = 0)$	C	D
C	1	−1
D	2	0

TABLE 2. Fitness with vengeance

$(v > 0)$	C	D
C	1	$-1-v$
D	$2 - v/c$	0

[5] See Fehr and Gaechter (1999).
[6] See, for example, Levine (1998).
[7] See Frank (1988) for a detailed discussion of this idea.

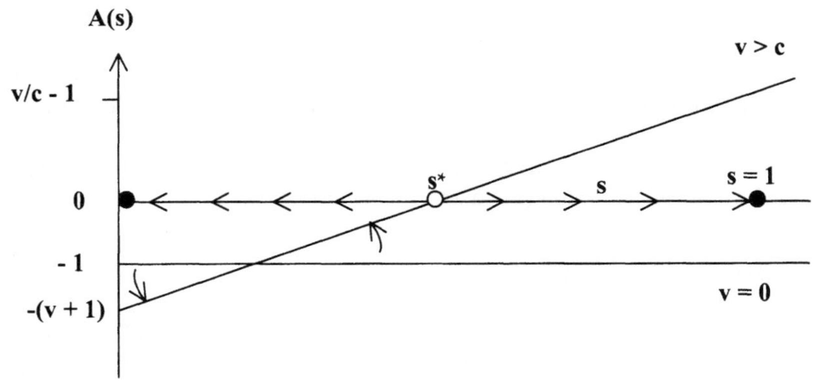

FIG. 1. The advantage of cooperating. The fitness advantage $A(s) = W(C) - W(D) = (v/c - 1)s - (v + 1)(1 - s)$ is graphed as a function of the population fraction s playing C for two values of the vengeance parameter v. The graph of A rotates counterclockwise as v increases. The solid dot at $s = 1$ represents a socially efficient, fully cooperative Nash equilibrium.

are equal at $s^* = (1 + 1/v)/(1 + 1/c)$. For $s < s^*$ the expected fitness is higher for D and play converges to the inefficient (fitness 0) all-D equilibrium, as in the basic game. However, for $s > s^*$ the expected fitness is higher for C, and play converges to the efficient all-C equilibrium.

Hence, for $v > c$ we have a coordination game with two locally stable, pure Nash equilibria and an unstable, mixed Nash equilibrium at $s^* < 1$ (Fig. 1). Thus the threat of vengeance can deter defection and support fully cooperative, socially efficient behavior (C,C) as a Nash equilibrium.

3 The Viability Problem

The vengeance motive v is itself subject to evolutionary forces, perhaps slower than those determining the prevalence s of cooperation, but real forces nonetheless. The expected fitness of a cooperator is $W(C|\ s, v) = 2s - 1 - v(1 - s)$, which is a strictly decreasing function of v for any fixed $s < 1$. Only when there are no culprits ($s = 1$) is the expected fitness independent of v. Thus the fitness of player v is weakly dominated by that of player v' whenever $0 < v' < v$. Assuming that players occasionally encounter culprits, the vengeance preference parameter v will be driven towards 0 under any plausible evolutionary dynamics. We thus have a second-order free-rider problem, and it seems that vengeance is not viable.

Standard solutions do not work well for this viability problem. Of course, the problem is attenuated for social creatures that form groups of closely related individuals, such as slime molds (relatedness near $r = 1$) or ants ($r = 2/3$), but we are interested in humans, whose groups typically consist of individuals who are not that closely related (say $r = 0.05$). If each individual's v were observable, then those with higher v might encounter D-play less frequently (as in Robert Frank's 1988 discussion) and thus maintain equal or higher fitness. This "greenbeard" solution[8] ignores the evolutionary pressure for lower-v individuals to mimic the visible signs of higher v. As noted below, building a reputation is a bit complicated in our setting. Less standard solutions are discussed at greater length in our related work[9] and include a continuing stream of mutants (Sethi and Somanathan 1996), and ruling out intermediate values of v (Axelrod 1986; Huck and Oechssler 1999) and moralistic strategies of extended negative reciprocity (Boyd and Richerson 1992).

4 Groups and Memes

During most of our evolutionary history, humans, like other social primates, presumably lived in small groups of individuals. In constructing our model, we assume that the typical person interacts every day with other members of his or her group, and also often interacts with others outside the group, but repeat encounters with any particular outsider are sporadic. In this setting, cooperation within the group can be maintained by many forms of reciprocation (positive or negative, direct or indirect), but cooperation outside the group is problematic: see Sect. 7 for a discussion.

All known groups of humans maintain memes that prescribe appropriate behavior toward fellow group members, and typically prescribe different appropriate behavior toward individuals outside the group.[10] The term "meme," as used here, seems to have been introduced by Dawkins (1976). He defines it (Dawkins 1982) as "the unit of information that is conveyed from one brain to another during cultural transmission." Durham (1991 pp. 188–190) illustrates the idea of memes with examples as diverse as marriage rules (which we would typically call social norms), and labeling boundaries for colors (at which wavelength of light does the color term change in different societies). Blackmore (1999) focuses on imitation as a mechanism for cultural transmission of memes but, as is implicit in our analysis, transmission mechanisms are richer than simple imitation.

The analysis below focuses on competition among memes that prescribe behavior toward culprits outside the group and toward group members who deviate from that prescription. These memes determine the group's reputation and therefore the fitness its members receive in encounters outside the group.

The success of the meme, as with any other adaptive unit, is measured by its ability to displace alternatives, i.e., by its fitness. There are many distinct mechanisms by which one meme may displace another, ranging from warfare to fashion, but for the most part these mechanisms align with the most fundamental mechanism, enhanced individual fitness. Without necessarily accepting assertions[11] that misalignments are always minor and temporary, our analysis will assume that a meme prescribing a par-

[8] This term is from Boyd and Richerson (1998), and is a fanciful but striking example of observability.
[9] See Friedman and Singh (1999).
[10] Sober and Wilson (1998) provide a catalogue of illustrations of this idea: we use some of their illustrations below.
[11] Wilson (1975) is a prominent proponent of this view.

ticular degree of vengeance is fitter than existing alternatives when it brings higher individual fitness to group members on average.[12]

Groups affect individual fitness in several ways. As already noted, they provide gains from internal cooperation and (depending on the reputation they carry) some gains from external cooperation. They also regulate access to scarce resources such as favorable home sites, stored food, and marriage partners. Status or prestige within the group affects access: Sober and Wilson (1998) extensively document these kinds of patterns of norms. For example, here is their illustration from Jewish society in thirteenth-century Spain (itself taken from MacDonald 1994):

> A law was then in place that a man who did not pay his taxes would have a blot placed on his genealogy . . . [A] marriage could not take place unless the genealogies of both the bride and the groom were consulted and approved. A blot on a person's genealogy could prevent a marriage. Thus, failure to pay taxes could have severe costs . . . Similar laws discouraged individuals from converting to other religions or even associating with members of other groups. (p. 145)

Another striking example provided by Sober and Wilson comes from Levine's (1965) study of Ethiopian culture: "The Amhara peasant who avenged an insult is 'following an ethic of cardinal importance'" (p. 182). Sober and Wilson discuss several other variants in norms across different societies.[13]

5 Model Elements

The model has two parameters describing relevant memes.

- $v^n \geq 0$ is the group's normative vengeance level, i.e., the prescribed cost that group members are supposed to incur when punishing outgroup culprits.
- $a \geq 0$ is the group's tolerance of deviations from that norm. A deviation $x = v^n - v$ incurs a status loss $x^2/(2a)$ and, of course, corresponding status gains by others.[14]

The relevant genes are summarized in one parameter.

- $v^{max} \geq 0$ is the maximum possible taste for vengeance that any meme could induce, given the individual's capacity for anger and his malleability.

Individual characteristics are described by:

- $v \in [0, v^{max}]$ is the actual vengeance cost an individual prefers; we assume it is learned from personal experience within the group, through fitness-maximizing best responses.

---

[12] Note that we are putting aside the issue of redistribution within the group in this formulation. There is no alternative meme within the group: selection is therefore essentially at the group level, because all those in the group have the same meme. For further discussion and numerical examples, see Sober and Wilson (1998).

[13] Their source is the Human Relations Area File, a database developed by anthropologists to facilitate comparisons across cultures. It contains coded and indexed ethnographies of hundreds of different cultures.

[14] This assumption can be relaxed, so that there is a net loss to the group. This generalization is outlined in the paragraph following the derivation of Result 1, in Sect. 7.

- $\bar{v} \in [0, v^{max}]$ is the group average degree of actual vengeance cost incurred in punishing outgroup culprits.

Finally, the impact of group reputation is captured in:

- $f(\bar{v}) = \exp(-\bar{v}/b)$ is the frequency with which an individual encounters culprits (who are all outgroup members, in equilibrium). It decreases in the group's reputation, summarized in $\bar{v}$, and increases in the positive parameter b, representing the hostility of the environment.

6 Results

6.1 *Result 1: The level of vengeance that maximizes the fitness of group members is* $v^o = \max\{0, b - 2\}$

This level would be advantageous for the group to induce in its members, given the hostility b of the environment. The intuition is straightforward. Under present assumptions, sanctions against deviators sum to zero within the group, so the group optimum simply trades off the increased fitness cost of higher v against the increased benefit of encountering fewer culprits. Section 7 shows that this tradeoff implies an optimum at v^o.

Of course, this result describes optimal rather than actual behavior, and ignores the second-order free-rider problem that no individual captures much of the benefit, which is dispersed throughout the group. To predict actual behavior, we now consider a fixed meme summarized by v^n and a, and examine how individual preferences will adapt.

6.2 *Result 2: Individual adaptation drives* v *and* $\bar{v}$ *toward the individual optimum* $v^* = v^n - a$*, truncated to* $[0, v^{max}]$

For given group characteristics $\bar{v}$ and v^n, the individual fitness cost is proportional to the sum of the direct cost v and the loss $(v^n - v)^2/(2a)$ from deviating from the group norm. The sum is minimized at $v^* = v^n - a$. Individual fitness has a single peak at v^*, so adaptive dynamics push the individual's parameter towards this optimum. The optimum is attained as long as the value is within the feasible range; otherwise v^* is truncated below at 0 and above at v^{max}. Assuming that individual adaptation is faster than either memetic or genetic evolution, v^* is a good approximation of the actual preferences of each individual, and an even better approximation of their average.

What is the relation between the group optimum v^o and the individual optimum v^*? The group meme, embodied in the parameters a and v^n, is subject to selective pressures in the medium run, and group average fitness is single-peaked. Any group whose memes bring actual behavior (near v^* by result 2) closer to the true optimum (v^o by result 1) has a selective advantage. Hence, memetic evolution will drive actual behavior toward the optimum, as long as it is feasible,[15] but if v^o is infeasible (because

[15] We do not formally model the dynamics of this process, but it is a straightforward exercise. The idea is that groups with memes that maximize group average fitness increase their popu-

the environment is so hostile that $v^o > v^{max}$), then there is a genetic selective advantage to increasing v^{max}.[16] Hence we have the following conclusion.

6.3 Result 3: Coevolution of memes and genes drives actual behavior v^* toward the group optimum v^o; in equilibrium $v^n = max\{0, a + b - 2\}$

Thus the actual vengeance level is socially efficient in evolutionary equilibrium. Note that the memes, v^n, that support this efficient behavior are not the optimum value v^o, but rather exaggerated versions, $v^n = v^o + a$.

On longer time scales, there can be shifts in the environment b and in the punishment technology c. These shifts affect the encounter function f, and hence the group optimum v^o. Our main conclusion implies that memes (and, when necessary, genes) will adjust under selective pressure so that individual behavior v^* will track the new group optimum.

7 Details and Generalizations

Here we explain how conclusions generalizing results 1–3 arise from less restrictive specifications than those used in the text, and spell out some details of the arguments. For concreteness, we assume that the underlying social dilemma is still as in Table 1.

Even without imposing a group structure, one can write the expected fitness advantage to cooperating $A(p, u) = W(C) - W(D)$ as a function of the individual's probability estimate p that his partner will choose C, and his expectation u of her vengeance parameter. One can derive p and u from a general specification of noisy observables, as discussed in Friedman and Singh (1999). Figure 2 shows how the fitter choice, C or D, depends on the sign of A, and how the choice shifts with the individual's own parameter v. One can derive[17] a smooth, decreasing encounter function $f(\bar{v})$ from the assumption that everyone chooses according to Fig. 2. One can also develop a theory of group size and the reliability of group reputation from similar considerations.[18] The text shortcut these matters by implicitly obtaining p and u from a convenient but arbitrary encounter function $f(\bar{v})$ together with a given (high) level s of within group cooperation.

The crucial feature of the group is that it imposes an expected fitness loss $\rho(x)$ when an individual deviates $x = v^n - v$ from the group norm v^n. The loss function ρ is

---

lation share at the expense of other groups. Sober and Wilson (1998) provide an example that they call a "smoking gun" for cultural group selection in action. Drawing on the work of Kelly (1985), they summarize how the Nuer, an African pastoral group, rapidly expanded at the expense of the neighboring Dinka group prior to European colonization (pp. 186–191).

[16] Roughly, the mechanism would be as follows. Individuals with higher levels of v^{max} would stand out in the group, and have higher status. This will affect reproductive success. Obviously increases in v^{max} take place on a longer time scale.

[17] See Friedman and Singh (1999).

[18] See Friedman and Singh (1999).

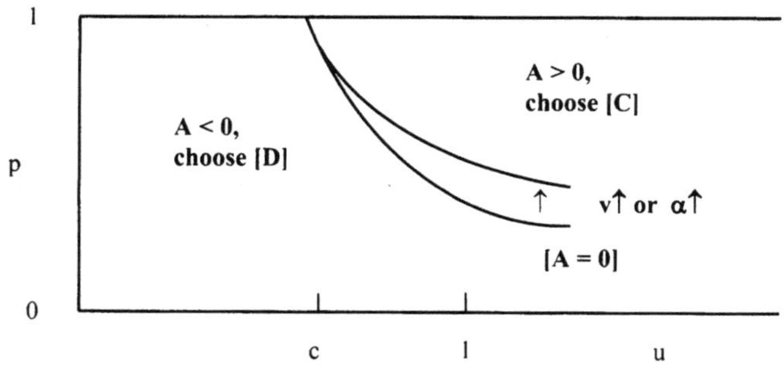

FIG. 2. The decision rule. The appropriate choice of C or D is given by the sign of the advantage function $A(p, u)$, where p is the probability that the partner will choose C, and u is an unbiased estimate of the partner's vengeance parameter. The $A = 0$ locus moves up with increases in the decision-maker's direct (v) or full ($\alpha = v + \rho(v^n - v)$) vengeance cost

assumed to be smooth, convex (with slope >1 for x sufficiently large), and minimized at 0, but it need not be quadratic or even symmetric.

7.1 Result 3: Remarks

The text uses a direct argument from earlier results to establish the striking result that memes solve the second-order free-rider problem and align actual behavior with optimal behavior. Here, we highlight two underlying assumptions. The first is that there is a hierarchy of time scales, so that individual v (and hence also $\bar{v}$) adapts most rapidly, then the memes v^n and/or a, then the genes v^{max}, and finally the environmental and technological parameters b and c. It does seem that v adapts rapidly to social memes; e.g., according to stories in the media, children raised in Belfast and Lebanon and brought to the US for the summer have no problem adapting within a few months to the low-v US norm, and then adapting back when they return.

It is natural to think of memes as evolving faster than genes, but that is not necessary for our result. Examples of coevolution on overlapping time scales include sickle-cell anemia in yam-growing areas, and lactose tolerance in herding communities.[19] The optimality result clearly fails when the environment or punishment technology changes faster than the memes (or, when v^{max} is too low, faster than the genes).

The other key assumption is that there is no fitness conflict between memes and genes, so memes are selected that bring higher fitness to group members. This assumption holds for memes transmitted vertically (e.g., from parents to children) because they share a common fate with genes and hence their fitnesses tend to coincide. However, horizontal meme transmission (e.g., from one teenager to another) need not respect genetic fitness. Here our assumption (and result 3) may or may not hold. Our reading of Sober and Wilson's wide-ranging discussion and other anthropological evidence supports both vertical and horizontal trans-

[19] These examples are developed in Durham (1991).

mission of v^n and a, but we have not seen any examples of gene and meme conflict.

7.2 Result 2: Derivation

With probability $f(\bar{v})$, a v-cooperator encounters a defector and receives fitness loss $(1 + v + \rho(v^n - v))$, i.e., the sucker payoff plus the cost of wreaking vengeance plus the group status loss from departing from the norm. The individual receives a fitness gain of 1 in an encounter with a cooperator. Thus the individual's expected fitness is $W(v \mid \bar{v}, v^n) = 1(1 - f(\bar{v})) - (1 + v + \rho(v^n - v))f(\bar{v}) + \mathrm{R} = 1 - f(\bar{v})(2 + v + \rho(v^n - v)) + \mathrm{R}$, where R is the base-level fitness including the (positive) effect on one's status from other group members' deviations from the norm v^n. The fitness function does not account for the possibility that the individual may sometimes play D, but this omission is harmless in terms of analyzing the adjustment of v.

Since the current choice of v has negligible effect on $\bar{v}$, the derivative of W is, up to the positive multiplicative constant $f(\bar{v})$, simply $\rho'(v^n - v) - 1$. The assumptions on ρ ensure that this derivative is decreasing in v and equal to zero at a unique $v^* > v^n$. Hence, the single-peaked property holds, and the argument in the text shows that individual adaptation pushes actual v towards v^*. With the quadratic specification for ρ we have the first-order condition $0 = \rho'(v^n - v) - 1 = (v^n - v)/a - 1$, so in this case $v^* = v^n - a$, truncated to $[0, v^{max}]$, as claimed.

7.3 Result 1: Derivation

The text assumes that the group imposes ρ purely through status changes. In this case, R cancels the mean contribution of ρ, so the group average fitness is $W^g(\bar{v}) = 1(1 - f(\bar{v})) - (1 + \bar{v})f(\bar{v}) = 1 - f(\bar{v})(2 + \bar{v})$. The group optimum v^o maximizes this expression on $(c, v^{max}]$. Using $f(v) = \exp(-v/b)$, the first-order condition reduces to $2 + v = -f/f' = b$, so v^o is $b - 2$, truncated to $(c, v^{max}]$ as claimed.

More generally, imposing a loss on a deviator may reduce the group's overall fitness by some fraction $t \in [0, 1]$; e.g., some potential gains to cooperation may not be realized. Group average fitness becomes $W^g(\bar{v}) = 1 - f(\bar{v})(2 + \bar{v} + t\rho(v^n - \bar{v}))$. With f and ρ as specified in the main text, direct computation yields $v^o = a(-t/2) + b(1 - t) - 2$. If $t = 0$, we have the case just analyzed, where v^o depends only on the environmental hostility parameter b. Higher t decreases the optimal level of vengeance and introduces positive dependence of v^o on the tolerance parameter a and the environmental hostility parameter b.

8 Discussion

Vengefulness, or a taste for negative reciprocity, is a crucial part of the human emotional repertoire. We model its importance in sustaining cooperative behavior, but highlight an intrinsic free-rider problem: the fitness benefits of vengeance are dispersed throughout the entire group, but the fitness costs are borne personally. Evolutionary forces tend to unravel people's willingness to bear the personal cost of punishing culprits. The countervailing force that sustains vengeance is a group norm together with low-powered (and low-cost) group enforcement of the norm. Such memes coevolve with personal tastes and capacities to produce the optimal level of vengeance.

The underlying interaction in our model is the simplest possible social dilemma, but our methods easily extend to more complex interactions. It seems straightforward to redefine a culprit as one who harms any group member, not necessarily oneself. More generally, if the interaction took the form of a common pool resource or public goods game, a player would be considered a culprit to the extent that his contribution falls short by an amount e from the efficient level (or a normative level). The utility bonus could then take the form $ve \ln h$, in which case the total harm to the culprit would be Ve/c, where V is the sum of the other participants' vengeance parameters. Results parallel to 1–3 above seem to follow.

Positive reciprocity could be analyzed directly in a similar fashion: preferences could include a utility gain (but no fitness gain) for rewarding a partner's cooperation, and a social norm could impose a fitness loss on deviators. However, since culprits are rare and cooperators are ubiquitous in a successful society, the fitness cost of the rewards is excessive when relying entirely on positive reciprocity to sustain cooperation. Negative reciprocity greatly reduces the burden.

We can speculate how our model would apply in different societies. The application to hunter–gatherer bands or villagers is clearest; here parameter b directly reflects the uncooperative tendencies of people from neighboring groups, and c reflects the opportunities to identify, track down, and inflict harm on them. In more highly structured societies, vengeance is often exacted by delegated specialists (the criminal justice system). The marginal cost c of vengeance is lower, but is still positive, so the model remains valid. Here an important shortcoming is that the model takes as exogenous the institutions that determine the vengeance technology.

What are the empirical implications of our model? Laboratory experiments can distinguish a taste for negative reciprocity from the egalitarian preferences recently hypothesized,[20] and generally favor our hypothesis.[21] New laboratory experiments can be conducted with a basic game similar to that in Table 1, together with a punishment technology similar to that used in the paper, $h = cv$. The punishment cost parameter c can be controlled at several levels. The impact of c on the frequency of punishment choices reveals the distribution of vengeance preferences v in a given subject population. Pilot experimental results, reported in Brownlow et al. (2000), indicate that many UCSC undergraduate subjects have large positive values of this preference parameter.

Our model also has clear comparative statics predictions that in principle are testable with anthropological data: norms of vengeance and vengeful behavior should vary systematically with the hostility of the environment, the technology for harming culprits, and the technology for enforcing group norms. If the model is on the right

[20] See, for example, Fehr and Gaechter (1999), Bolton and Ockenfels (1999), and Fehr and Schmidt (1999).

[21] See also, Falk et al. (1999).

track, one can hope that dysfunctional vengeful behavior (as in the Balkans) might improve in coming decades as the relevant memes evolve.

Acknowledgments. The first author is grateful to CES Munich University for hospitality while writing the first fragments in May 1997. We have benefited from the comments of Robert Boyd and Jack Hirshleifer, and seminar audiences at JAFEE2000, Indiana, Purdue, UCLA, and UCSC. We are especially grateful to Yuji Aruka for his helpful advice and suggestions. Remaining shortcomings are our responsibility.

References

Axelrod R (1986) An evolutionary approach to norms. Am Polit Sci Rev 80:1095–1111

Blackmore S (1999) The meme machine. Oxford University Press, Oxford

Bolton GE, Ockenfels A (1999) ERC: a theory of equity, reciprocity and fairness. Smeal College of Business Administration, Pennsylvania State University, http://silmaril.smeal.psu.edu: 80/lems/ERC.pdf

Boyd R, Richerson PJ (1985) Culture and the evolutionary process. University of Chicago Press, Chicago

Boyd R, Richerson PJ (1992) Punishment allows the evolution of cooperation (or anything else) in sizable groups. Ethol Sociobiol 13:171–195

Boyd R, Richerson PJ (1998) The evolution of human ultra-sociality. In: Eibl-Eibisfeldt I, Salter FK (eds) Indoctrinability, ideology and warfare: evolutionary perspectives. Berghahn, New York

Brownlow A, Gullapalli R, Hoskins B, Ross M (2000) Game theory: introducing vengeance—a laboratory experiment. UC Santa Cruz, Master's class term project

Dawkins R (1976) The selfish gene. Oxford University Press, New York

Dawkins R (1982) The extended phenotype: the gene as the unit of selection. Freeman, San Francisco

Durham WH (1991) Coevolution: genes, culture, and human diversity. Stanford University Press, Stanford

Falk A, Fehr E, Fischbacher U (1999) On the nature of fair behavior. Institute for Empirical Economic Research, University of Zurich, http://www.unizh.ch/iew/wp

Fehr E, Gaechter S (1999) Cooperation and punishment in public goods experiments. Institute for Empirical Economic Research, University of Zurich, http://www.unizh.ch/iew/wp

Fehr E, Schmidt K (1999) A theory of fairness, competition, and cooperation. Q J Econ 114:817–868

Frank R (1988) Passions within reason: the strategic role of the emotions. Norton, New York

Friedman D, Singh N (1999) On the viability of vengeance. UC Santa Cruz Working Paper, http://zzyx.ucsc.edu/Econ/workpapers.html

Hagen WW (1999) The Balkans' lethal nationalisms. Foreign Aff 78:52–64

Huck S, Oechssler J (1999) The indirect evolutionary approach to explaining fair allocations. Games Econ Behav 28:13–24

Kelly R (1985) The Nuer conquest: the structure and development of an expansionist system. University of Michigan Press, Ann Arbor

Levine DK (1998) Modeling altruism and spitefulness in experiments. Rev Econ Dyn 1:593–622

Levine DN (1965) Wax and gold: tradition and innovation in Ethiopian culture. University of Chicago Press, Chicago

MacDonald KB (1994) A people that shall dwell alone: Judaism as a group evolutionary strategy. Praeger, Westport
Nowak MA, Sigmund K (1998) Evolution of indirect reciprocity by image scoring. Nature 393:573–577
Sethi R, Somanathan E (1996) The evolution of social norms in common property resource use. Am Econ Rev 86:766–788
Sober E, Wilson DS (1998) Onto others: the evolution and psychology of unselfish behavior. Harvard University Press, Cambridge
Wilson EO (1975) Sociobiology: the new synthesis. Belknap Press of Harvard University Press, Cambridge

Avatamsaka Game Structure and Experiment on the Web

YUJI ARUKA

Summary. *Avatamsaka* is a well known *Mahayana Buddhist Sutra*. A Japanese professor, working in the field of Buddhist philosophy, skillfully illustrated the situation of heaven and hell in terms of the Avatamsaka (Kamata 1988 pp. 167–168). Suppose that two people sit down at a table, across from each other. They are bound with rope so that one arm only is free, and are then each given a very long spoon. This spoon is so long that they cannot feed themselves with it. There is enough food for both of them on the table. If they cooperate and feed each other, they will both be happy. This is defined as heaven. However, if the first is kind enough to provide the second with a meal, but the second does not feel cooperative, then only the second gains. This must give rise to a feeling of hate in the first. This situation denotes hell. The gain structure does not only depend on an altruistic willingness to cooperate. On an individual level, there is no difference between cooperation and refusal, and the same is true for risk taking. A situation of expected maximization of utility gives infinite equilibria. Our interest is to find a way to heaven from the other possible situations. This chapters is concerned with how an actual player would react in such experiments.

Key words. Avatamsaka, Positive spillovers, Evolutionary game, Experiment on the Web

1 Some Features of the Avatamsaka Game

1.1 Road Example

We first set the Avatamsaka game in a realistic example of a secular situation, such as the road example. We imagine driving a car in a busy place. When we wish to enter a main street from a side road or a parking area along the road, it would be much better for us if someone gave way for us, although that person would not gain anything by giving way to someone they might never meet. This kindness, done by a stranger, is our gain. The symmetrically opposite situation in which we give way to a stranger shows the same relationship. Anybody who does a kind deed for someone can never be guaranteed to gain from that other person. If nobody ever gave way to anybody

---

Faculty of Commerce, Chuo University, Higashinakano, Hachioji, Tokyo 192-0393, Japan

TABLE 1. Coordination game and Avatamsaka game

Strategies	Player B	
(i) Coordination game		
Player A	Strategy 1	Strategy 2
Strategy 1	(a, a)	(b, c)
Strategy 2	(c, b)	(d, d)
(ii) Avatamsaka game		
Player A	Defection	Cooperation
Defection	(0, 0)	(1, 0)
Cooperation	(0, 1)	(1, 1)

else, the situation on the roads would be attrocious. This supports the old axiom, "One good turn deserves another."[1]

It seems to me, however, that we usually adopt a kind of mixed strategy rather than a pure strategy. Perhaps we rarely behave as extremists in any situation. This may be the result of our own *cultural climate*, which develops as a consequence of the actions of agents living in the community.

1.2 Two-Person Game Forms

Here, the heaven and hell in Avatamsaka are shown in a numerical example. The Avatamsaka game is a kind of *coordination game*. This can be seen from the situation in Table 1, where

$$a = 0,\ b = 1,\ c = 0,\ d = 1$$

This can easily be checked by considering one of the hell situations: if you choose "cooperation" but the other player chooses "defect," you lose your point. Put another way, you can only get a point by choosing defect, if the other player is altruistic.

1.3 Infinite Equilibria under Expected Utility Maximization

We first state briefly how this game gives infinite equilibria when utility maximization is expected. To do this we remove any equilibrium-selection mechanism from our framework. Under the previous conditions it was important to reach a multiple equilibrium with all other players. The present problem unambiguously exits from this argument. The expected player A who always uses strategy D, when player B uses strategy D with probability q, as well as strategy C with probability $1 - q$, is

$$E(\mathrm{D}, q) = q \cdot 0 + (1 - q) \cdot 1 = 1 - q$$

For the case when player A always uses strategy C, their expected utility is

[1] This axiom is deeply meaningful, and should not be illustrated in terms of an individualistic utility function.

$$E(C, q) = q \cdot 0 + (1 - q) \cdot 1 = 1 - q$$

It can be seen that any mixture of strategies can be called a *mixed strategy* if all the expectations happen to be equal. There are indeed an infinite number of solutions q fulfilling

$$E(C, q) = E(D, q)$$

It is clear that the welfare results of this game are well ordered in a configuration of complete coordination, i.e., a combination of (cooperation, cooperation) as the top level, to a lower combination as the decisions to adopt cooperation decrease if a repeat game is introduced.[2]

1.4 No Complementarity to be Found Except for Positive Spillovers

In our game we cannot find any complementarity relation, as is often seen in other coordination games. A game is *complementary*[3] if a higher action in terms of general welfare by player B increases the marginal return from an action of player A. The situation can then be given a coordination feature. First define a marginal return-of-action switch. If player A can switch their action from strategy 1 to strategy 2 when player B's action is strategy 1, the presence of an action switch can give the marginal return

$$\Delta^{B:1}_{A:1\text{ to }2} = c - a > 0$$

If player B adopted strategy 2, the same switch may, for example, give the higher marginal return[4]

$$\Delta^{B:2}_{A:1\text{ to }2}(A) > \Delta^{B:1}_{A:1\text{ to }2}(A)$$

The welfare arrangements of choice combinations are strictly ordered from higher to lower:

$$(d, d) \text{ is preferred to } (a, a)$$

In our Avatamsaka game, there may not be any complementarity relations[5]

$$\Delta^{B:2}_{A:1\text{ to }2}(A) = \Delta^{B:1}_{A:1\text{ to }2}(A)$$

On the other hand, the coordination game usually has another property called *positive spillovers*. This property implies that each player can increase their own reward

[2] This payoff matrix is

$$P = \begin{pmatrix} 0 & 1 \\ 0 & 1 \end{pmatrix}, \quad P^t = \begin{pmatrix} 0 & 0 \\ 1 & 1 \end{pmatrix}$$

It can be seen that the rank of the matrix is 1, i.e., either the rows or the columns are linearly dependent. The solution of the game then *degenerates*.

[3] This holds, for instance, if $b < a < d < c$. For definitions of complementarity and spillover see, for example, Cooper (1999).

[4] $d - b > c - a > 0$ since $\Delta^{B:2}_{A:1\text{ to }2} = d - b$.

[5] $d - b = 1 - 1 = 0$; $c - a = 0 - 0 = 0$.

as a result of the other player's switch of action. In our setting, each player can increase their reward by the other player switching from strategy 1 to strategy 2:

$$\Delta_{A:1}^{B:2\text{ to }1}(A) = b - a > 0, \qquad \Delta_{A:2}^{B:2\text{ to }1}(A) = d - c > 0$$

Positive spillovers exist in Avatamsaka:[6]

$$\Delta_{A:1}^{B:1\text{ to }2}(A) = 1, \qquad \Delta_{A:2}^{B:1\text{ to }2}(A) = 1$$

Thus, it can be seen that there are no complementarities, but there are positive spillovers, in our Avatamsaka game.

1.5 How to Win the Game

In my teaching class experiences, I have noticed that many students hold a standard which is some way away from supposed rational economic behavior: a *double standard* in the sense that there is a double binding action of the classical principle and the other one. We show a few examples. One is where students give the wrong answer when the exercise asks them to decide how to impose a penalty to prevent a rival defecting in a *repeated* prisoners dilemma game. In this game,

$$c < a < d < b$$

Students often favor the penalty of

$$(b - d) + (d - c) = b - c$$

They should choose an answer which gives a positive result for themselves. However, they strongly dislike the idea that the difference $b - c$ between the players will increase, although they are only asked for a deterrent to the defection of the other player. There are few students who impose the penalty of

$$(b - d)$$

as *a deterrent* to a rival's defection. Students or other subjects never stick to the classical principle of payoff maximization. Instead, they are adherents of the principle of winning the game.[7]

It is clear that there are many cases which do not conform to the expected utility maximization principle hypothesis. The above example shows that coordination cannot be guaranteed even though a complementary relation exists. "How to win the game is often irrelevant to the payoff maximization principle." Players therefore maneuver some kind of spiteful action to attempt to maximize their *shadow points*, for example, they "choose an action which will increase the difference in points between themselves and the other player as quickly as possible, and at almost every possible opportunity." It can immediately be seen that this difference describes the negative spillovers for player B in our terminology:

$$c - d$$

[6] $b - a = 1 - 0 = 1$; $d - c = 1 - 0 = 1$.

[7] See Taylor (1985), Chap. 5, where such a discrimination principle is referred to. Another example from experimental results can be found in Cason et al. (1997).

In this context, our game may, in the first stage, result in the solution (defect, defect), because this solution can give the largest difference between players:

$$d - c = 1 - 0 = 1$$

However, this only produces all losers, and nobody gains anything, which contradicts the "winning" principle and results in a "draw." Thus, the winning principle in this game needs to be more sophisticated. In our game, it seems very difficult for players to employ the winning principle, although they can use some kind of spiteful action.

1.6 Conversion of the Avatamsaka Game into a Coordination Game

Some economists who are familiar with the orthodox way of thinking may wonder whether this Avatamsaka structure deserves investigation by economists. They might suggest a transformation of the Avatamsaka structure into the more usual coordination one. In order to achieve this, it is only necessary to introduce perturbation δ as a very small number.

A perturbed version of case (i) (Table 2) may satisfy a complementary relation in a weaker form:

$$\Delta^{B:2}_{A:1\text{ to }2}(A) = d - b = 1 - 1 = 0$$

$$\Delta^{B:1}_{A:1\text{ to }2}(A) = c - a = 0 - \delta = -\delta$$

Hence

$$\Delta^{B:2}_{A:1\text{ to }2}(A) > \Delta^{B:1}_{A:1\text{ to }2}(A)$$

Another way to introduce perturbation may be considered, as in case (ii) (Table 2). In this case, it also follows[8] that

$$\Delta^{B:2}_{A:1\text{ to }2}(A) > \Delta^{B:1}_{A:1\text{ to }2}(A)$$

TABLE 2. Perturbed Avatamsaka games

Strategies	Player B	
Case (i)		
Player A	Defection	Cooperation
Defect	$(0 + \delta, 0 + \delta)$	(1, 0)
Cooperation	(0, 1)	(1, 1)
Case (ii)		
Player A	Defection	Cooperation
Defect	(0, 0)	$(1-\delta, 0)$
Cooperation	$(0, 1-\delta)$	(1, 1)

[8] $\Delta^{B:2}_{A:1\text{ to }2}(A) = d - b = 1 - (1 - \delta) = \delta$ and $\Delta^{B:1}_{A:1\text{to}2}(A) = c - a = 0 - 0 = 0$.

These perturbed versions may create some trivial cases. This configuration may only give a dominant strategy for defect in case (i) or for cooperation in case (ii). No mixed strategy can be constructed in a meaningful way.[9]

1.7 The Trembling Hands Example

We now turn to another perturbed version of the game which may be likened to *trembling hands* as caused by *emotional problems*. On the one hand, a player may feel vexed when they offer cooperation and the other player defects. Again, we presume that this is a symmetrical situation. On the other hand, a player may feel pleased when they defect if the other player cooperates. We treat the feelings of vexation and pleasure symmetrically by the use of a small amount, δ, which may be identified with the pleasurable feeling if δ is positive, and by $-\delta$ if the feeling is identified with vexation. We assume that they have a common absolute value.[10] We now set $\delta > 0$, and thus we can consider the next proposition.

Proposition: Suppose our payoff matrix for the perturbed version with human emotions is

$$P = \begin{pmatrix} 0-\delta & 1+\delta \\ 0 & 1 \end{pmatrix}$$

Players must employ the mixed strategy to use either cooperation or defection, with probability 0.5, whatever number is taken for δ.

Proof: Suppose player A employs the strategies $p = (p_1, p_2)$, and player B employs $q = (q_1, q_2)$. The value of this game on the payoff matrix P is then

$$v = q_2(p_2 + p_1(\delta + 1)) - p_1 q_1 \delta$$

Substituting $1 - p_2, 1 - q_2$ for p_1, q_1, we get

$$v = (p_2 - 1)\delta + q_2(1 - 2(p_2 - 1)\delta)$$

Player B's choice to maximize player A's expectation is required to be[11]

$$q_2 = 0.5$$

This also holds for player A. Henceforth, either player can use the strategy probability 0.5 to maximize his expected utility. QED

This proposition may only suggest a technical possibility of a way out of our dilemma. It must not be overlooked that the emotional reasons might often be a means of settlement.[12]

---

[9] In case (i), for example, $E(\mathrm{D}, q) = q \cdot \delta + (1 - q) \cdot 1 = 1 - (1 - \delta)q$ and $E(\mathrm{C}, q) = q \cdot 0 + (1 - q) \cdot 1 = 1 - q$ together would require $\delta = 1$ if a mixed strategy did exist.

[10] A utility function incorporating psychological relations is treated in Taylor (1985), Chap. 5.

[11] Solve $\frac{dv}{dp_2} = 0$, given q. Since $\delta - 2q_2\delta = 0$, it follows that $q_2 = 0.5$. Similarly, $\frac{dv}{dq_2} = 0$ can be solved given p. There is a mixed strategy where both result are different.

[12] In a different context, Frank (1988) argued for the virtues of human emotion. He said that emotions are especially useful for providing a fairly accurate measure of a person's intentions,

2 Evolutionary Dynamics in our Game

2.1 A Payoff Table for Group Encounters

In the above proposition, we have shown why a player can use this *mutant type* of mixed strategy, which lies between pure defection and pure cooperation. Furthermore, it has the remarkable property of being unaffected by the magnitude of δ. This mutant, when he encounters the representative mutant, as it lies between pure defection and pure cooperation, can at most have an expected utility of 0.5. Now assign, for example, a probability

$$(\text{defect, cooperation}) = (0.8, 0.2)$$

It then holds that

$$0.8(0.5 \cdot 0 + 0.5 \cdot 1) + 0.2(0.5 \cdot 0 + 0.5 \cdot 1) = 0.5$$

which is equivalent to our representative mutant

$$0.5(0.5 \cdot 0 + 0.5 \cdot 1) + 0.5(0.5 \cdot 0 + 0.5 \cdot 1) = 0.5$$

There is a possibly infinite number of combination between pure defection and pure cooperation, but neither of them could gain more than our mutant strategy. I therefore believe that such behavior, i.e., to employ this even probability at each strategy, could be representative of all other mixed-type strategies, as will be seen at the end of this paragraph. Even if this strategy predominated, the set of such strategies could not dominate the situation with the best solution (cooperation, cooperation), whose expected utility or welfare is always set at 1.

In an *evolutionary game*, the higher the average welfare of the community can be, the greater the growth rate of the group. It is interesting to speculate whether or not a generation of mutant mixed strategies would evolve into an increasingly dominant influence in the community. The payoff structure of our evolutionary game, or a table of the payoff outcome for group encounters, is summarized in Table 3.[13]

This table shows the remarkable property that all the groups have the same payoff. It is natural that every group should have expected to survive, and also that they should believe in their own rationale. The initial states may all be unchanged, and in this case, any *shuffle* action would not be useful.

Finally, it must be noted that this idea should be still kept, even if some variety of mixed strategies is to be introduced. Suppose there are three types of mixed strategy.

because emotional responses are generally beyond a person's purposeful control, making them difficult to fake consistently and rendering a person's actions automatic and consistent.

[13] Defect vs. defect implies that the resultant points for the D-group become 0 with probability 1, namely, $1 \cdot 0 \rightarrow 0$. Defect vs. cooperation implies that the resultant points for the D-group become 1 with probability 1, namely, $1 \cdot 1 \rightarrow 1$. Defect vs. mutant implies that the resultant points for the D-group become 0.5, since the player has defected with probability 0.5 and cooperated with probability 0.5, namely, $0.5 \cdot 0 + 0.5 \cdot 1 \rightarrow 0.5$. Similarly it follows for the C-group that cooperation vs. defect implies $1 \cdot 0 \rightarrow 0$. Cooperation vs. cooperation implies $1 \cdot 1 \rightarrow 1$. Cooperation vs. mutant implies $0.5 \cdot 0 + 0.5 \cdot 1 \rightarrow 0.5$. It also follows for the mutant group that Mutant vs. defect implies $0.5 \cdot 0 + 0.5 \cdot 0 \rightarrow 0$. Mutant vs. cooperation implies $0.5 \cdot 1 + 0.5 \cdot 1 \rightarrow 1$. Mutant vs. mutant implies $0.5 \cdot 0 + 0.5 \cdot 1 \rightarrow 0.5$.

TABLE 3. Payoff structure of evolutionary game

Species	D-group	C-group	Mutant
D-group	0	1	0.5
C-group	0	1	0.5
Mutant	0	1	0.5

TABLE 4. Payoff relationship of revised evolutionary game

Species	D-group	C-group	Mutant
D-group	$-\delta$	$1 + \delta$	0.5
C-group	0	$1 + \delta$	0.5
Mutant	0	1	$0.5 + \delta$

We can, for example, classify three groups according to the magnitude of the probability that they will employ D. If a player employs D with a probability of 0.8 *on average*, they belong to the D-biased group. If this probability is 0.5, they belong to the neutral group, and if it is 0.2, they belong to the C-biased group. This player then has the same payoff relationship with any group.[14] This situation does not generate any motion because these interactions take place without any perturbation. We now consider the even probability case of a mutant as a *representative* of mixed strategy groups.

2.2 Modified Tables for Evolutionary Dynamics with a Reinforcement Mechanism

Table 3 only gives relationships of complete dependency between groups, since they all depend on the initial values of their positions, and this does not change once the initial values have been implemented. However, there is the possibility of motion in the *psychological reaction* of a player to the defection or cooperation of his opponent. This hypothesis can give some remarkably interesting dynamics. Taking this into account, we now consider the situation given in Table 4.

Note that δ is only a *psychological factor*. The total points gained never contain an additional δ. However, δ implies a *reinforcement* for the player to continue to employ a strategy. A premium δ may encourage a player either to continue using their last action, or to change to the current action. We can design this game so that each player is able to observe the list of all the current points of the other players. It is assumed that players can only decide their own strategy by taking account of the total points of others, i.e., the community average without any premium. There may also be a case which is so sophisticated that every player can decide their strategy by calculating the community average including a premium.

[14] If a different player employs a D-biased strategy, he expects to earn $0.8(0.8 \cdot 0 + 0.2 \cdot 1) + 0.2(0.8 \cdot 0 + 0.2 \cdot 1) = 0.2$ when he is faced with a member of the same D-biased group, and so on.

We now formulate *a population dynamics with a reinforcement mechanism.* Let the payoff matrix of the community interaction be A, its row be a_i, and each proportion of the community population employing each strategy be x_i, where x_1 represents the ratio of the defection group to the community population, is x_2 the ratio of the co-operation group, and x_3 is the ratio of the mutant group. Denote

$$x = \begin{pmatrix} x_1 \\ x_2 \\ x_3 \end{pmatrix}$$

and let its transpose be x^t. Here, $\Sigma_{i=1x_i}^{3}=1$.

There are three ways to describe the community interaction matrix. Any player who employs D may feel either vexed when defeated or pleased when winning.[15] The D-group then is faced with the following matrix:

$$A^{\mathrm{D}} = \begin{pmatrix} -\delta & 1+\delta & 0.5 \\ 0 & 1 & 0.5 \\ 0 & 1 & 0.5 \end{pmatrix}, \qquad a_1^{\mathrm{D}} = (-\delta, 1+\delta, 0.5)$$

Any player employing C may be reinforced by their own feelings about premium δ when they adopt C against the same group. The C-group matrix is

$$A^{\mathrm{C}} = \begin{pmatrix} 0 & 1 & 0.5 \\ 0 & 1+\delta & 0.5 \\ 0 & 1 & 0.5 \end{pmatrix}, \qquad a_2^{\mathrm{C}} = (0, 1+\delta, 0.5)$$

Similarly, a player employing M may be reinforced by their own feeling of premium δ when adopting M against the same group. The M-group matrix is

$$A^{\mathrm{M}} = \begin{pmatrix} 0 & 0 & 0.5 \\ 0 & 1 & 0.5 \\ 0 & 1 & 0.5+\delta \end{pmatrix}, \qquad a_3^{\mathrm{M}} = (0, 1, 0.5+\delta)$$

The welfare of a group whose players decide the group's strategy can then be expressed as $a_i^i\ A^i\ x$, and the average welfare as $x_i^i\ Ax$, where $a_i^i\ x$ is the i-th group's scores or welfare, and $\frac{\Delta x_i}{x_i}$ is the i-th group's population growth rate. Our simple dynamics show that a relative population of the group increases if the group welfare is greater than the community average welfare. Thus, our evolutionary dynamics equations can be stated as[16]

$$\Delta x_1 = x_1(a_1^{\mathrm{D}} x - x^t A x)$$
$$\Delta x_2 = x_2(a_2^{\mathrm{C}} x - x^t A x)$$
$$\Delta x_3 = x_3(a_3^{\mathrm{M}} x - x^t A x)$$

---

[15] When a player is faced with a mutant player whose strategy is the even probability policy, the premium may be offset, on average, as zero.

[16] Alternatively, we may use a more sophisticated system by replacing A with A^i. A computer run of this sophisticated system $a_x^i x - x^t A_x^i$ may produce essentially similar results because it shares the same cumulating points.

2.3 A Motive for Metamorphosis

Group population dynamics work through an adjustment mechanism such that the group with the highest growth rate whose score is higher than the average community score can attract players from the other groups, or in other words, induce them to change from their original strategy to the strategy of the group with highest score. The group population dynamics alone does not represent any individual willingness, and the change is based on the expected utility maximization.

We assume that each player has an equal probability of encountering each group, and the probability that the i-th player will meet one particular group depends on the current scores of the groups. The motive for a change in action is mainly related to how large scores are distributed among groups. A group formation can be caused by a kind of metamorphosis, for example, from C to D, and C to M; from D to C, and D to M; from M to C, and M to D.

Two forces for dynamic motion can therefore be seen in our revised population dynamics. The C-group is attracted into the D-group if the D-group's *current* score is greater than that of the C-group. The *marginal preferences* of the groups may be ordered hypothetically according to the current distribution of the proportions of the groups in the community, i.e.,

$$(x_1, x_2, x_3)$$

On the other hand, the i-th group must reinforce itself if its profit is above average, i.e.,

$$a_i^i x - x^t Ax$$

Thus, every product of

$$x_i(a_i^i x - x^t Ax)$$

may make an instantaneous net difference between the exit or entry of each group, i.e., *metamorphosis*. We call a psychological factor *asymmetrical* if D has a payoff of $-\delta$ against D. Furthermore, if D also has a payoff of $1 + \delta$ against C, we call the factor *symmetrical*.

When psychological effects are introduced, they have a *complementary relationship*, as defined in the previous section, whether they are asymmetrical or symmetrical.

2.4 Population Dynamics in a Triangle

In our dynamics, we can always discover whether a mutant group will eventually disappear or dominate. Extreme points in the unit simplex of probability support show the domination of a single group, with the two extreme points of the base being two groups trying to remove the other group. This dominant group stands on a base defined by the D-group at one extreme point and the C-group at the other.

Theorem:[17] The strategies $x = (x_1, x_2, x_3)$, representing the ratios of D, C, and M, are all in a unit simplex. A player M employs their mutant strategy randomly with even

[17] The proof of this theorem can be seen in Aruka (2000).

probability. Suppose that there is a psychological complementary relation characterized by δ in the above definition.

1. When the metamorphosis from M to C works well, and if the initial starting point is chosen appropriately, there may be a guaranteed path toward an increase in the C ratio and a fixed point, with M being exhausted.
2. There may be a large enough increase in the magnitude of the psychological factor δ to reverse the metamorphosis from C to M.
3. The simultaneous increase in x_1 and x_2, coevolution of C and D, come from the metamorphosis of M.
4. The extremes D and M cannot be attracting points.

Let the *Liapunov function* be

$$V(x) = x_1^{p1} x_2^{p2} x_3^{p3}$$

It then follows that

$$\dot{V}(x) = V(x)\left(p_1 \frac{\dot{x}_1}{x_1} + p_2 \frac{\dot{x}_2}{x_2} p_1 \frac{\dot{x}_3}{x_3} \right)$$

If $\dot{V}(x) > 0$ is always confirmed, then any vector x will asymptotically converge to a fixed point $x^* = (p_1, p_2, p_3)$ (Arrow-Smith and Place 1982 pp. 207–210).

We now straightforwardly illustrate these statements in Figs. 1–3. In the system $\{a_i^t x - x^t A x\}$, Figure 1 shows a phase diagram of dynamic motion based on psychologically asymmetrical matrices, while Fig. 2 is a phase diagram based on psychologically symmetrical matrices. Figure 3 in the system $\{a_i^t x - x^t A^i x\}$ shows a convergent trajectory towards pure C, provided that the order of magnitude of δ has been changed from 0.01 to 0.1.

FIG. 1. Asymmetrical case of psychology

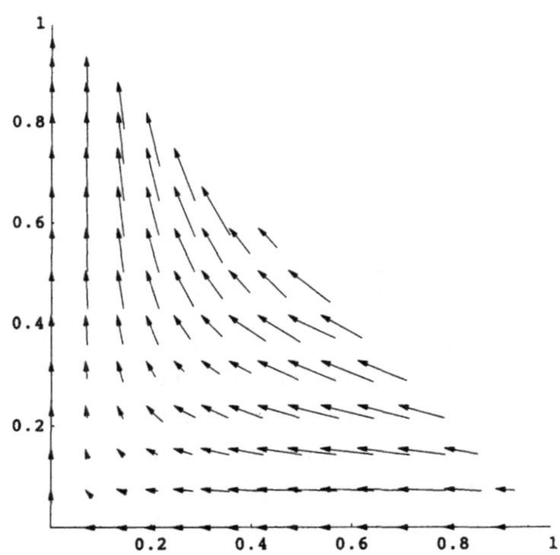

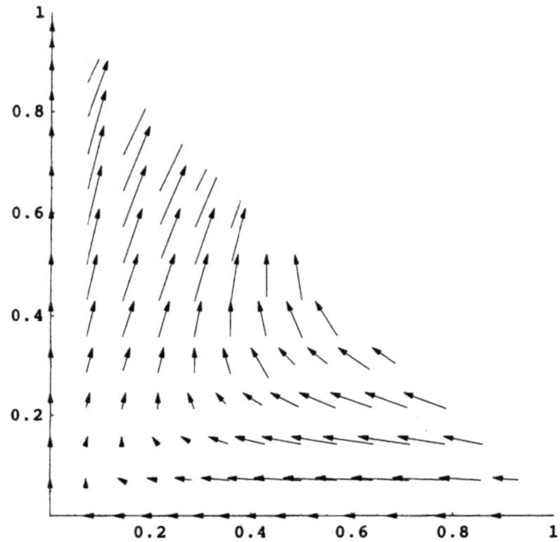

FIG. 2. Symmetrical case of psychology

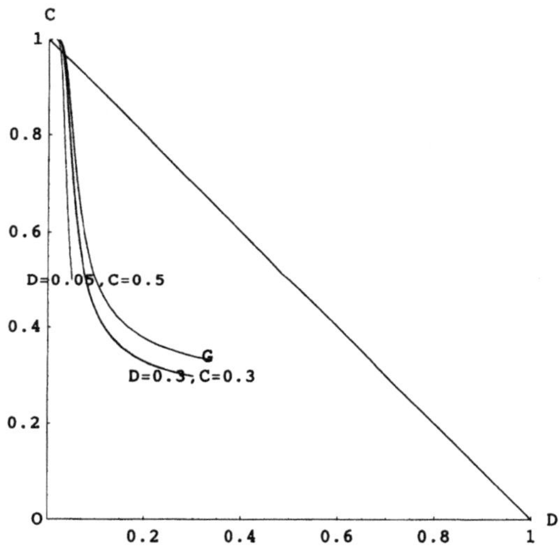

FIG. 3. Pure-C forming

2.5 Effect Generated by an Intelligent Group

We end this section by considering the statement that the *Volterra–Lotka relation does not hold.* For the D-group to survive as a *winner* presupposes at least the positive ratio of the C-group. Then pure C, or always using strategy C, will eventually die out. For the C-group to survive, it does not need the other group since it is achieving the best outcome for the community. Considering D vis a vis C, D is a *predator* while C is the *prey*: a predator–prey relationship seems to hold. If this relation lasted for ever, the growth rates of the two species would give a closed orbit, as in the Volterra–Lotka equation. However, there are many differences between this supposition and ours. *The difference here is that a predator may temporarily alter itself into prey; in other words,*

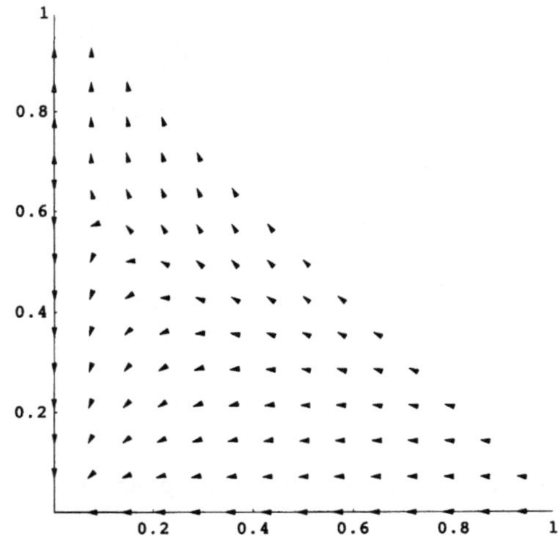

FIG. 4. Intelligent case

D becomes C for a while. This is why our game will not give regular fluctuations in the group ratios, and much more complicated behavior may be anticipated.

We have used a mutant strategy in the sense that a player employs either D or C *randomly*. Any discretion by using an *intelligent* strategy was never presumed. Now suppose that a very tactically skillful player penetrates an opponent's strategy and gains over the average of the community. Such a player must occasionally use a policy of "tit for tat,"[18] but their personal ability never propagates among others unless a *coalition group* organized by the smart player is permissible. Such an intelligent mutant will not realistically survive. So far in our game, skillful conjecture did not matter in the M-group. Suppose that a skillful player has a payoff matrix as in Table 5. Notice that a psychological score of "over $\frac{1}{2}$" must be realizable. The opponents who are facing the intelligent group must face a loss in their payoff. In details, it follows that

$$A^{D} = \begin{pmatrix} -\delta & 1 & 0.5-\delta \\ 0 & 1 & 0.5-\delta \\ 0 & 1 & 0.5+\delta \end{pmatrix}, \quad A^{C} = \begin{pmatrix} 0 & 1 & 0.5-\delta \\ 0 & 1+\delta & 0.5-\delta \\ 0 & 1 & 0.5+\delta \end{pmatrix}, \quad A^{M} = \begin{pmatrix} 0 & 1 & 0.5-\delta \\ 0 & 1 & 0.5-\delta \\ 0 & 1 & 0.5+\delta \end{pmatrix}$$

If the intelligent player should form their own coalition or group to attract other players, it would be easy to show that this group's rapid increased would soon reach critical proportions. See Fig. 4. There would be different cases in the final state depend-

TABLE 5. Payoff for a skillful man

Species	D-group	C-group	Mutant
Mutant	0	1	over $\frac{1}{2}$

[18] Even in this case, the player cannot keep 1 point, because their behavior might induce someone else to resist and exert D against them.

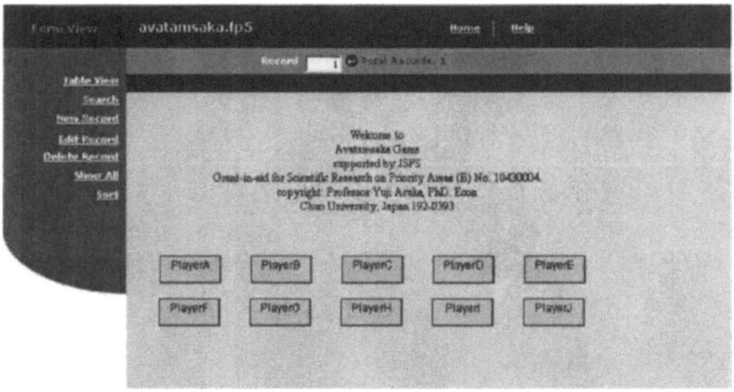

FIG. 5. Cover page of experimental program

ing on the initial conditions, e.g., (1) possibly all the mutants in the population, or (2) all the C-group, and so on.

Our *hypothetical* population dynamics could result in a *limiting sequence* toward a special base, or an extreme point in whatever direction it turned. Some types of *degeneration* have been observed in our computer runs. It is interesting to consider how, and under what conditions, an experiment done with human subjects would show a similar result. *The aim of this research, however, was to test statistically the data on the ratio of groups obtained by experiments with various different samples.*

3 Experiment on the Web

3.1 Experiment on the Web

This experiment had a maximum of ten subjects, as is standard practice in experimental economics (Fig. 5). The experiment may be found on the Web by using the software program *File Maker Pro.*[19] A near equivalent of the display layouts for this experiment, which were developed by myself, will be reproduced on the Web by a well-functioning CGI loaded in File Maker. It may not be important to a user on the Web what operating system is employed. However, it should be noted that only my server operates with the use of File Maker, and a maximum of ten subjects can easily and simultaneously access my address via an internet browser.[20] Note that File Maker Pro

[19] The advantage of database software File Maker Pro ver.5 is that "Web Companion," a very useful function of CGI (common gateway interface), is loaded.

[20] There are possible combinations with an even number total of 2, 4, 6, 8, or 10, since our experiment is designed for a two-person game. Two persons are required for each match. If there are ten subjects, then one must play a match with the remaining nine persons. We must therefore arrange a total of matches $(10/2) \times (10 - 1) = 45$. The game is then finished. Incidentally, our

is not essential. A maximum of ten personal computers are sufficient if each has an internet browser installed, such as Internet Explorer 4.0 or a later version, to support CSS1 (Cascading Style Sheet 1). Unfortunately, Netscape Navigator is not suitable at present.

3.2 Evolution of Strategies

In the computer simulation and in the experiment with human subjects, our aim was to investigate how three groups, such as a pure defection group, a pure cooperation group, and a mutant group {D, C, M}, as dealt with in the previous sections, would fluctuate. Different sequences were used in each sample in the experiment, in order to examine what degree of variance was due to different samples. Such samples may be domestic or international.

3.3 Monitoring the Others

Following convention, one exercise was given to the participants at first. We confined the series of games to a maximum of ten subjects for ten successive times. Every subject could have as much information as their rivals. They can always monitor the current points of the other players, as well as their names. This partly takes into account the assumption of learning, which is discussed below.

3.4 Two Kinds of Experiment

In this experiment, we are also interested in checking whether a *pecuniary motive* would affect a player's action or not. Two different games, with and without payment, had to be designed. Experiments in economics are usually used to test the pecuniary motives of economic behavior. Thus an experimental economist can generally reward each player's efforts. In our game, we wish to control the experiment rather differently. We therefore arrange two types of experiment. One experiment follows the traditional method of experimental economics of paying each player according to their achievement. The other is without payment. Our intention is to test statistically whether the during motive in our game was biased toward a pecuniary advantage or not.

3.5 Repeated Games

Our experiment needs ten subjects and an instructor. Each subject is requested to play a two-person game within the rules which appear in the computer program (Fig. 6). One game may consist of a series of matches to be played with the other players. When there are ten subjects, each game consists of nine matches for each player. The player can run their game by referring to the points distribution of the whole community. This points distribution may also be reproduced in each game display. The players will repeat the same game ten times. Consequently, each player will play 90 matches.

---

program is also valid for all the standard two-person games such as the prisoner's dilemma simply by revising the coefficient in the payoff matrix.

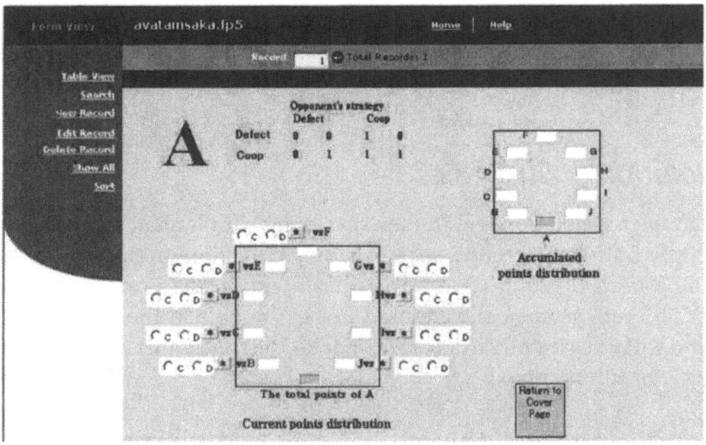

FIG. 6. Display for player A

The maximum possible number of points a player can receive is 90. On the other hand, the worst possible number may be 0.[21]

3.6 Our Initial Concerns

In our original Avatamsaka game scheme without the introduction of perturbation, only *positive spillovers* were implemented. The individual's motives, whether egoistic or altruistic, did not work well owning to the entire lack of a *complementary relation* in the payoff structure. We were concerned whether or not *positive spillovers* isolated from complementarities could still induce behavior based on utility maximization.

In an iterated game, *learning effects* are one of the most important and controversial topics. It is usually assumed that without any repetition, the subjects cannot know the distribution of types of players in advance. By iterations of the same game, the subjects should learn this distribution. Even if there could be an assured *mathematically* symmetrical payoff matrix, by which a unique solution is guaranteed, nevertheless, it would take much too long, possibly an infinitely long time, to know the true effects of the learning thus experimentally revealed. Gale (1960) pointed this out over 40 years ago, when he gave a classical account of learning in a two-person game.[22] Fortunately, our simulation results show that some sequences reach a stationary situation in a finite number of steps. We therefore restricted the number of trials in

[21] Instructor. A computer expert will be necessary to assist students to operate Internet Explorer or a similar browser. Subject must give comment. A final request will be made in the following form: "Please give us a comment on the Avatamsaka game after finishing this game. Please use this line to enter the mail-sending form or equivalently (questionnaire) the left frame. This comment will automatically be sent by the mail form system written in CGI."

[22] A convergence element might be too slow to deserve practical consideration, even if convergence were guaranteed. See Gale (1960) pp. 250–256.

our game to ten, and these experiments will of course be shorter than those in the computer simulation. However, if some tendency toward an eventual situation is to be discernible, the data observed must be statistically tested.

Furthermore, there have been many investigations into whether or not an agent really learns to behave in a way which will increase their payoff. Testing learning effects is still an active area of experimental economics. However, we do not especially want to focus on any stochastic process of convergence by learning, and/or to ask whether a human being is rational or not. These are not our concerns here. Instead, we are interested in the question: *How can an emotion to cooperate be generated in concert?* In fact, however, we have introduced a kind of learning process. In order to achieve a well-organized experiment to examine this problem, we have uncovered a range of types by informing the subject of the gains of others. If the whole composition of the ratios of the groups were always monitored by each player, and at least partly detected by them knowing the current scores of the other players, they could each react unambiguously according to their own deductions, which are closely

TABLE 6. Avatamsaka experiment in December 2000, in Germany (points and C-ratio)

	A	B	C	D	E	F	G	H	I	J
(i) Points distribution										
Round 1	9	6	7	6	8	7	7	7	6	7
Round 2	6	7	4	5	7	6	7	7	7	6
Round 3	7	7	5	5	7	7	6	5	5	6
Round 4	7	6	7	8	7	8	7	7	9	6
Round 5	9	8	8	7	8	8	7	7	8	7
Round 6	9	9	7	8	8	6	6	8	8	7
Round 7	8	3	6	6	7	7	4	8	8	6
Round 8	8	4	6	5	8	7	6	8	8	5
Round 9	8	8	7	7	8	6	6	3	7	6
Round 10	5	5	5	4	6	4	4	5	4	5
Total	76	63	62	61	74	66	60	65	70	61
Average	0.84	0.7	0.69	0.68	0.82	0.73	0.67	0.72	0.78	0.68
(ii) Times that strategy C was employed										
Round 1	8	8	9	3	7	7	8	9	4	6
Round 2	8	8	9	2	8	5	6	0	4	9
Round 3	6^{NA_2}	8	9	6	7	7	6	0	5	6
Round 4	9	8	9	7	9	7	8	3	7	5^{NA_1}
Round 5	9	8	9	8	8	5^{NA_1}	5	9	9	7
Round 6	9	8	9	6	8	8	3	9	8	7
Round 7	9	8	9	6	8	8	7	7	9	6
Round 8	8	8	7	6	8	8	6	0	8	6
Round 9	8	8	6	0	9	8	5	9	8	4
Round 10	9	8	7	0	0	4	4	0	9	5
Total	83	80	83	44	72	67	58	46	71	61
Average	0.92	0.89	0.92	0.49	0.80	0.74	0.64	0.51	0.79	0.68

The superscript ^{NAx} shows that the number of *NA* (not available) values is *x*.

related to their emotional feelings. In our experiment, the emphasis is placed on taking into consideration the community trend, but not on individual learning by collecting information.

3.7 Experiment in Frankfurt am Main in December 2000

Finally, we mention the latest results of our Avatamsaka experiment held in Germany.[23] These are given in Table 6, but with no statistical tests or any evaluation of their consequences.

Acknowledgments. The project underlying this paper was financially supported by Grant-in-Aid for Scientific Research (B) (1) No. 1043004 of the Japan Science Promotion Society. Also partly supported by Nomura Foundation for Social Science. The first draft was written in Clare Hall, Cambridge, UK, in May 2000. This paper was presented at the J.-W. Goethe Universitaet for Frankfurter Volkswirtschaftliches Kolloquium (Frankfurt Economic Lectures) on November 27, 2000. I am especially grateful to Prof. Bertram Schefold for providing me with his class and facilities to conduct the experiment of December 12, 2000. Also special thanks to Prof. Stephen Gaustello, Marquette University, USA, who has sequentially organized our experiment since October 2000, and much contributed to improve our experimental design.

References

Arrow-Smith DK, Place CM (1982) Ordinary differential equations. Chapman & Hall, London

Aruka Y (2000) Avatamsaka game structure and its design of experiment. Working paper No. 3, Institute of Business Research, Chuo University

Cason T, Saigon T, Yamato T (1997) Voluntary participation and spite in public good provision experiments: international compasion. Mimeo

Cooper RW (1999) Coordination games. Cambridge University Press, Cambridge

Frank R (1988) Passions within reason: the strategic role of emotions. Norton, New York

Gale D (1960) The theory of linear economic models. University of Chicago Press, Chicago, (reprinted)

Kamata S (1988) Kegon no shiso (The thought of Avatamsaka) (in Japanese). Kodan-sha, Tokyo

Taylor M (1985) Possibility of cooperation. Cambridge University Press, Cambridge

[23] This experiment on the Web was held at the PC Pool, J.-W. Goethe Universitaet (Frankfurt University), Frankfurt am Main, from 1100 to 1200 hours, on December 12, 2000. The subjects were supplied from the class of Prof. Bertram Schefold, Faculty of Economics, and Roland Goslar, his assistant, who conducted the experiment by accessing the author's computer server in Tokyo, Japan. This experiment belongs to a class without explicit payment, although non-pecuniary prizes of cookie were allocated among subjects according to their resultant scores.

Part IV Multiagent Systems and Complexity

On the Relevance of Genetic Programming to Evolutionary Economics

Shu-Heng Chen

Summary. In this chapter, we review the development of artificial adaptive economic agents in evolutionary economics. The review starts from a 1986 paper by Robert Lucas, a Nobel Prize laureate in economics. From there, we shall see how the idea of economic adaptive agents was enriched and implemented by Holland's two books on genetic algorithms (Holland 1975) and on classifier systems (Holland et al. 1986). We then examine the impact of Holland's artificial adaptive agents on two different groups of economists. One was led by Thomas Sargent, representing new classical economics, and the other by Brian Arthur, standing for Santa Fe Institute (SFI) economics. A moot point here is that the spirit of the genetic algorithm (GA) (John Holland's legacy) is lost in mainstream economics, but is reserved in SFI economics. We then move to Koza's genetic programming, and show how John Holland's legacy was further expanded in evolutionary economics.

Key words. Artificial adaptive economic agents, Building blocks, Genetic programming, LISP S-expression, Evolving populations of decision rules

1 Motivation

While genetic programming has been applied to economic modeling for more than half a decade, its relevance to the nature of economics has not been fully acknowledged. In the most sympathetic situations, it is regarded as nothing but *alchemy*. In unsympathetic situations, it is notorious for its *black-box* operation. Sometimes, the processes and results are so complicated that economists rarely consider it relevant or interesting. This chapter is not an attempt to make economists embrace genetic programming, but from a scientific viewpoint, we would like to avoid prejudice due to a lack of solid understanding. This chapter is intended to deliver a simple but strong message: *genetic programming is not just another fancy technique exploited by the unorthodox, but could be a faithful language to express the essence of economics.* In

AI-ECON Research Center, Department of Economics, National Chengchi University, Taipei 11623, Taiwan particular, it provides evolutionary economists with a way to *substantiate* some features which distinguish them from mainstream economists.

To achieve this goal, I will address the following list of questions:

- Why is genetic programming relevant to economics?
- Are genetic operators relevant to economics?
- Technically speaking, what makes economists like or dislike genetic programming?

2 An Evolving Population of Decision Rules

Let us start from the most fundamental issue: *why is genetic programming relevant*? Lucas (1986) provided the notion of an economic agent. "In general terms, we view or model an individual as *a collection of decision rules* (rules that dictate the action to be taken in given situations) and *a set of preferences* used to evaluate the outcomes arising from particular situation-action combinations." (Lucas 1986 p. 217; italics added.) Immediately after the *static description* of the economic agent, Lucas added an *adaptive* (*evolutionary*) version of it. "These decision rules are continuously under review and revision: new decision rules are tried and tested against experience, and rules that produce desirable outcomes supplant those that do not." (ibid; p. 217). So, according to Lucas, the essence of an economic agent is *a collection of decision rules which are adapting (evolving) based on a set of preferences*. In brief, it is the idea of an *evolving population*.

Supposing that an evolving population is the essence of the economic agent, it then seems important to know whether we economists know any operational procedures to substantiate this essence. Back in 1986, the answer was absolutely *no*. That certainly does not mean that we did not know anything about evolving one *decision rule*. On the contrary, since the late 1970s, the literature on what is known as the bounded rationality in macroeconomics has introduced a number of techniques to evolve a single decision rule (a single equation or a single system of equations): recursive regression, Kalman filtering, and Bayesian updating, to name a few. Sargent (1993) made an extensive survey of this subject. However, these techniques shed little light on how to build a Lucasian agent, especially since what we wanted to evolve was not a single decision rule but a population of decision rules.

In fact, it may sound a little surprising that economists in those days rarely considered an individual as a population of decision rules, never mind attending to the details of its evolution. Therefore, all the basic issues pertaining to models of the evolving population received little, if any, attention. For example, how does the agent *initialize* a population of decision rules? Once the agent has a population of decision rules, which one should they follow? Furthermore, in what ways should this population of decision rules "be continuously under review and revision"? Should we review and revise them one by one because they are independent, or modify them together because they may correlate with each other? Moreover, if there are some "new decision rules to be tried," how do we generate (or find) these new rules? What are the relations between these new rules and the old ones? Finally, it is also not clear how "rules that produce desirable outcomes should supplant those that do not."

There is one way to explain why economists are not interested in, and hence not good at, dealing with a population of decision rules: economists used to derive the decision rule for the agent *deductively*, and the deductive approach usually leads to only one solution (decision rule), which is the *optimal* one. There was simply no need for a population of decision rules.

3 Genetic Algorithms and Classifier Systems

We do not know exactly when or how the idea of the *evolving population of decision rules* began to attract economists, but John Holland's contribution to *genetic algorithms* definitely exerted a great influence. Genetic algorithms simulate the biological evolution of a society of computer programs, each of which is represented by a chromosome or, normally, a string of binary ones and zeros. Each of these computer programs can be matched to a solution to a problem. This structure provides us with an operational procedure of the Lucasian agent. First, a collection of decision rules are now represented by a society of computer programs (a society of strings of binary ones and zeros). Second, the review and revision process is implemented as a process of natural selection. Table 1 shows how close the relation is between the Lucasian agent and the genetic algorithm.

While Holland's genetic algorithms have had a great impact on computer science, mathematics, and engineering since the early 1980s, their implications for social sciences were not acknowledged until the late 1980s. In 1987, Robert Axelrod, a political scientist at the University of Michigan, published the first application of the GA to the social sciences. A year later, the first PhD dissertation that applied the GA to social sciences was completed by John Miller from, not surprisingly, the University of Michigan. The issue addressed by Axelrod and Miller is the well-known *repeated prisoner's dilemma.* In addition to these two early publications, perhaps the most notable event that brought GAs into economics was the invited speech by Holland given at an economic conference at the Santa Fe Institute in the autumn of 1987. Among the audience were some of the most prestigious contemporary economists, including Kenneth Arrow, Thomas Sargent, Hollis Chenery, Jose Scheinkman, and Brian Arthur. In his lecture entitled "The global economy as an adaptive process," Holland introduced to the economics circle the essence of genetic algorithms: *building blocks.*

A building block refers to the specific pattern of a chromosome, i.e., an essential characteristic of a decision rule. There is a formal word for this in the genetic algorithm; it is called a *schema*. In the genetic algorithm, a schema is regarded as the basic unit of learning, evolution, and adaptation. Each decision rule can be defined as a combination of some schemata. The review and revision process of decision rules is

TABLE 1. The Lucasian agent and the genetic algorithm

Lucasian Agents	Genetic Algorithms
Decision rule	A string of binary ones and zeros
A collection of decision rules	A collection of binary strings
Decision rule review	Fitness evaluation
Good rules supplanting bad rules	Selection
Generation of new rules	Crossover and mutation
No equivalent	Schema

nothing more than a search for the right combination of those, possibly infinite, schemata. To rephrase Lucas's description in Holland's words, *economic agents are constantly revising and rearranging their building blocks as they gain experience*. Not only do genetic algorithms make the Lucasian economic agent implementable, but they also enrich its details.

After a gradual spread and accumulation of knowledge about the GA among economists, modeling economic agents with an evolving population of decision rules finally began to increase in the 1990s. To the best of my knowledge, the first referred journal article is Marimon et al. (1990). This paper is a follow-up of research by Kiyotaki and Wright (1989). In a simple barter economy, Kiyotaki and Wright found that low storage costs *are not* the only reason why individuals use money. The other one is that money makes it easier to find a suitable partner. Replacing the rational agents in the Kiyotaki–Wright environment with *artificially intelligent* agents, Marimon et al., however, found that goods with low storage costs play the dominating role as a medium of exchange.

The population of decision rules used to model each agent is a *classifier system*, another contribution made by Holland in the late 1970s. A classifier system is similar to the Newell–Simon-type expert system, which is a population of *if–then* or *condition–action* rules. However, the classical expert system is not adaptive. What Holland did with the classifier system was to apply the idea of competition in the market economy to a society of if–then rules. To implement market-like competition, a formal algorithm known as the *bucket-brigade algorithm*, credits rules generating good outcomes and debits rules generating bad outcomes. This accounting system is further used to resolve conflicts among rules. The shortcoming of the classifier system is that it cannot automatically generate or delete rules. Nonetheless, by adding the genetic algorithm on top of the bucket brigade and the rule-based system, one can come up with something similar to the Lucasian agent, who not only learns from experience, but can be *spontaneous* and *creative*.

While Holland's version of the adaptive agent is much richer and more implementable than the Lucasian economic agent, and the work was already completed before the publication of Holland's second book (Holland et al., 1987), its formal introduction to economists came 5 years after the publication of Lucas (1986). In 1991, Holland and Miller published a sketch of the *artificial adaptive agent* in the highly influential journal *American Economic Review*. The first technique to implement the Lucasian economic agent was finally "registered" in economic science, and genetic algorithms and classifier systems were formally added to the toolkit of economic analysis. Is 5 years too long? *Maybe not*, given that "Economic analysis has largely avoided questions about the way in which economic agents make choices when confronted by a perpetually novel and evolving world" (Holland and Miller 1991 p. 365).

What is next? If the Lucasian economic agent is a desirable incarnation of the economic agent in economic theory, and if Holland's artificial adaptive agent is indeed an effective implementation of it, then the follow-up research can proceed in three directions: first, the novel applications of this new technology, second, the theoretical justifications of the new technology, and finally, the technical improvement of it. That is exactly what we experienced during the 1990s.

For the first line of research, Jasmina Arifovic, a student of Sargent's, finished the first PhD dissertation that applied GAs to macroeconomics in 1991. It was not until

1994, however, that she published her work as a journal article. Arifovic (1994) replaced the rational representative firm in the cobweb model with Holland's adaptive firms, and demonstrated how the adaptation of firms, driven by market forces (natural selection), collectively make the market price converge to the rational-expectations equilibrium price. Since then, a series of her papers has been published in various journals with a range of new application areas, including inflation (Arifovic 1995), exchange rates (Arifovic 1996), and coordination games (Arifovic and Eaton 1995).

4 The SFI Economics

Although Holland introduced this powerful toolkit to economists, he did not conduct any economic research with this toolkit himself except for a joint work with Brian Arthur. Holland and Arthur met in September 1987 at a physics and economics workshop hosted by the Santa Fe Institute. They had a great conversation on the nature of economics. The *chess* analogy proposed by Arthur led Holland to believe that the real problem with economics is "*how do we make a science out of imperfectly smart agents exploring their way into an essentially infinite space of possibilities?*" (Waldrop 1992 p. 151). On the other hand, Arthur was impressed by Holland's approach to complex adaptive systems. Holland's ideas of adaptation, emergence, and perpetual novelty, along with other notions, offered illuminating revelations to Arthur, insights he could never have had gained if he had confined himself to theorizing on equilibria.

This new vision of economics turned out to be the approach of the Santa Fe Institute when it established its economics program in 1988. The essence of the SFI economics was well documented by Arthur (1992). Instead of explaining genetic algorithms and classifier systems, which Holland and Miller (1991) had already done, this paper put a great emphasis on motivation. Arthur eloquently argued why the deductive approach should give way to the inductive approach when we are dealing with a model of heterogeneous agents. His paper thus built the microfoundation of economics upon agents' cognitive processes such as pattern recognition, concept formation, and hypothesis formulation and refutation. Arthur then showed how the dynamics of these cognitive processes can be amenable to analysis with Holland's toolkit.

Maybe the best project to exemplify the SFI approach to economics is the *artificial stock market*. This research project started in 1988. Despite progress made in 1989, journal articles documenting this research were not available until 1994. Palmer et al. (1994) first built their stock market from a standard asset pricing model (Grossman and Stiglitz 1980). They then replaced the rational representative agent in the model with Holland's artificial adaptive agents, and then simulated the market. The three steps they took, i.e.,

- choosing a benchmark,
- replacing the rational representative agent with artificial adaptive agents,
- simulating the economy,

are the same as those in the procedure followed by Marimon et al. (1990) and Arifovic (1994), but unlike these two papers, the SFI has a quite different motive behind using the idea of artificial adaptive agents.

First of all, for new-classical economists, using artificial adaptive agents is mainly a way of understanding how the economy will converge to the rational-expectations equilibrium (REE) even when agents are not perfectly rational. Second, in a case where there are multiple equilibria, this device can be further used to select one of them. Third, when the situations get so complicated that the REE is not analytically available, this device can be used to "compute" the REE. Sargent (1993), in the first economics textbook to introduce genetic algorithms and classifier systems, gave a clear account of all these three advantages.

The SFI economists, in contrast, find these advantages rather minor. As Arthur (1992) stated:

> Yet it is hard to find much justification for why economics should concern itself deeply with learning and adaptive behavior. The new literature convinces us that learning models are useful in spelling out processes of adjustment to the standard equilibria of economic theory; and it raises fascinating questions about what it means to recursively incorporate new information into decision behavior. But it also leaves us with a vague feeling that learning is somehow ancillary to economics—an add-on adjustment dynamics to the core theory, not fully necessary perhaps to theorizing in the field though interesting in its own right. (p. 1)

There is little doubt that adaptive behavior can lead to the REE, but to the SFI economists, this can happen only if the problem is simple enough. For Arthur, the relevance of genetic algorithms to economics is much more than just strengthening the rational expectations equilibrium. He would like to see how one can use this tool to simulate the evolution of a real economy, such as the emergence of barter trading, money, a central bank, labor unions, and even Communists. However, he understood that one should start with a more modest problem than building a whole artificial economy, and this led to *the artificial stock market.*

Given this different motive, it is also interesting to see how the new classical economists and the SFI economists programmed agents in their models, and, given their coding or programming, how complex their agents can evolve to be. Arifovic (1994) simply coded a decision made by firms, namely, the quantity supplied, which is just a *number*. She did not code the *decision rules* of firms. Thus, generation after generation, firms were just crunching numbers. Nothing more complex or sophisticated can be expected from these firms no matter how long they last. Marimon et al. (1990) were a little different. They coded the decision rules of traders as exchange and consumption classifiers. The condition–action space is rather simple: there are only 72 possible classifiers in total; one can simply enumerate all of them. With such a limited variation, one can hardly expect any interesting complex behavior to evolve from this style of application.

Palmer et al. (1994) also used the standard trinary string to code different types of trading rules frequently used by financial market traders. Each bit of a string was randomly drawn from the trinary alphabet {0, 1, *}. Each bit corresponds to the *condition part* of a single trading rule. For example, the condition part of a *double moving average rule* could be "The 20-period moving average of price is above the 100-period moving average." The appropriate bit is 1 if the condition is true, and 0 if it is false. They typically used strings of 70–80 symbols, i.e., the same as the number of trading rules. This defines a search space with 3^{70} to 3^{80} possible nonredundant classifiers. However, each artificial trader has "only" 60 classifiers in their own classifier system.

Consider a case with 100 computerized traders: there are at most 6000 different rules being evaluated in one single trading run. Compared with the size of the search space, the number of rules is infinitesimal. This rather large search space is certainly beyond what Arthur (1992) called the *problem complex boundary*, a boundary beyond which arriving at the deductive solution and calculating it are unlikely or impossible for human agents, and this is where the SFI stock market comes into play. It provides the right place to use genetic algorithms, and a great opportunity to watch evolution. As depicted by Arthur (1992):

> We find no evidence that market behavior ever settles down; the population of predictors continually coevolves. One way to test this is to take agents out of the system and inject them in again later on. If market behavior is stationary they should be able to do as well in the future as they are doing today. But we find that when we "freeze" a successful agent's predictors early on and inject the agent into the system much later, the formerly successful agent is now a dinosaur. His predictions are unadapted and perform poorly. The system has changed. From our vantage point looking in, the market—the "only game in town" on our computer—looks much the same. But internally it coevolves and changes and transforms. It never settles. (p. 24)

Maybe the real issue is not whether GA are used to strengthen the idea of REE, or to simulate artificial life, but *how we program adaptive agents.* This is crucial because different programming schemes may lead to different results. As Frank Hahn pointed out, while there is only one way to be perfectly rational, there are an infinite number of ways to be partially rational (Waldrop 1992 pp. 250–251). This unlimited "degree of freedom" of programming adaptive agents was also noticed by Sargent (1993): "This area is wilderness because the researcher faces so many choices after he decides to forgo the discipline provided by equilibrium theorizing." (p. 2) Arthur would consider letting the agents start off *perfectly stupid*, and get *smarter and smarter* as they learn from experience. Now, comes the core of the issue: *how to program agents so that they can be initialized as perfectly stupid individuals, but can potentially get very smart.* To answer this question, let us go back to the origin of genetic algorithms.

5 Genetic Programming

It is interesting to note that the binary strings initiated by Holland were originally motivated by an analogy to machine codes. After decoding, they can be computer programs written in a specific language, say LISP or FORTRAN. Therefore, when a GA is used to evolve a population of binary strings, it behaves as if it is used to evolve a population of computer programs. If a decision rule is explicit enough not to cause any confusion in implementation, then one should be able to write it in a computer program. It is the *population of computer programs* (or their machine codes) which provides the most general representation of the *population of decision rules.* However, the equivalence between computer programs and machine codes *breaks down* when what is coded is the parameters of decision rules rather than decision rules (programs) themselves, as we often see in economic applications with GAs. The original meaning of evolving binary strings as evolving computer programs is lost.

The gradual loss of the original function of GAs has finally been noticed by John Koza. In 1987, when John Koza was on a London stopover after attending an AI conference in Italy, a friend handed him the proceedings of a conference on genetic algorithms. Impressed with the variety of applications, he nonetheless noticed a glaring omission: there was a lack of ways to *generate programs*. He then began to work on a system to breed *computer programs* genetically. He chose the language LISP as the medium for the programs created by genetic programming (GP) because the syntax of LISP allows computer programs to be manipulated easily like the bitstrings in GAs, so that the same genetic operations used on bitstrings in GAs can also be applied to GP. His breakthrough, however, "was deciding to identify the *units of crossover* not as single characters, or even as lines in a computer program, but as symbolic expressions (S-expressions) written in the LISP syntax." (Levy 1992 pp. 176–177; italics added).

5.1 LISP S-expressions

LISP S-expressions consist of either *atoms* or *lists*. Atoms are either members of a *terminal set*, that comprise the data (e.g., constants and variables) to be used in the computer programs, or they are members of a *function* set that consists of a number of prespecified functions or operators that are capable of processing any data value from the terminal set *and* any data value that results from the application of any function or operator in the function set. Lists are collections of atoms or lists, grouped within parentheses. In the LISP language, everything is expressed in terms of operators operating on some operands. The operator appears as the left-most element in the parentheses, and is followed by its operands and a closing (right) parenthesis. For example, the S-expression $(+X3)$ consists of three atoms: from the left-most to right-most they are the function "+," the variable X, and the constant 3. As another example, $(\times X(-Y3))$ consists of two atoms and a list. The two atoms are the function "×" and the variable "X," which is then followed by the list $(-Y3)$.

LISP was invented in the late 1950s by John McCarthy at MIT as a formalism for reasoning about the use of certain kinds of logical expressions, called recursion equations. LISP possesses unique features that make it an excellent medium for complex compositions of functions of various types, handling hierarchies, recursion, logical functions, self-modifying computer programs, self-executing computer programs, iterations, and structures whose size and shapes are dynamically determined. The most significant of these features is the fact that LISP descriptions of processes (routines) can themselves be represented and manipulated as LISP data (subroutines). As Koza (1992a) demonstrated, LISP's flexibility in handling procedures as data makes it one of the most convenient language in existence for exploring the idea of evolving computer programs genetically, however, Koza and others have noted that the use of LISP is not necessary for genetic programming; what is important for genetic programming is the implementation of a LISP–like environment, where individual expressions can be manipulated like data, and are immediately executable.

5.2 Symbolic Regression

The distinguishing feature of GP is manifested by its first type of application in economics, known as *symbolic regression*. In symbolic regression, GP is used to discover the underlying data-generation process of a series of observations. While this type of application is well known to econometricians, the perspective from GP is novel. As Koza (1992b) stated,

> An important problem in economics is finding the mathematical relationship between the empirically observed variables measuring a system. In many conventional modeling techniques, one necessarily begins by selecting the size and shape of the model. After making this choice, one usually then tries to find the values of certain coefficients required by the particular model so as to achieve the best fit between the observed data and the model. But, in many cases, *the most important issue is the size and shape of the model itself.* (p. 57; italics added)

Econometricians offer no general solution to the determination of size and shape (the functional form), but for Koza, finding the functional form of the model can be viewed as *searching a space of possible computer programs* for the particular computer program which produces the desired output for given inputs.

Koza employed GP to rediscover some basic physical laws from experimental data, for example, Kepler's third law and Ohm's law (Koza 1992a). He then also applied it to eliciting a very fundamental economic law, namely, the *quantity theory of money* or the *exchange equation* (Koza 1992b). Genetic programming was thus formally demonstrated as a *knowledge discovery* tool. This was probably the closest step ever made toward the original motivation of John Holland's invention: "Instead of trying to write your programs to perform a task you don't quite know how to do, *evolve them*." Indeed, Koza did not evolve the parameters of an arbitrary chosen equation; instead, he evolved the whole equation from scratch. This style of application provides an evolutionary determination of bounded rationality.

Koza (1992b) motivated a series of economic applications of genetic programming in the mid-1990s. Chen and Yeh (1996a) applied genetic programming to rediscovering *the efficient market hypothesis* in a financial time series. Chen and Yeh (1997a) then moved one step forward to propose an alternative formulation of the efficient market hypothesis in the spirit of the *Kolmogorov complexity* of algorithms for pattern extraction from asset price data. Chen and Yeh (1997a) and Szpiro (1997a) employed GP to discover the underlying chaotic laws of motion of time series data. Neely et al. (1997) and Allen and Karjalainen (1999) also adopted a GP approach to discover profitable technical trading rules for the foreign exchange market and the stock market, respectively. Another area in which GP was actively applied is *option pricing*. Chen et al. (1999) used GP for hedging derivative securities. Keber (1999) showed that genetically determined formulas outperformed most frequently quoted analytical approximations in calculating the implied volatility based on the Black–Scholes model. Chidambaran et al. (2000a) and Keber (2000) derived approximations for calculating option prices, and showed that GP-models outperformed various other models presented in the literature.

Needless to say, in the future, one can expect many more applications of GP to the automatic discovery of economic and financial knowledge (automatic generation of economic and financial knowledge in terms of their computer-programmed representations). However, its significant contribution to economics should not be mistaken for a perfect solution to knowledge discovery, data mining, or, more generally, *function optimization*. In a nutshell, genetic programming should be used to grow *evolving hierarchies* of building blocks (subroutines), the basic units of learning and information, from an immense space of subroutines. All evolution can do is look for improvements, not perfection. John Holland believed that these evolving hierarchies are generic in adaptation, and can play a key role in understanding human learning and adaptive processes. Hence, the value of GP to economists can best be seen as a model of human learning. From this standpoint, one may ask how GP, as a model of human learning, can enrich our understanding of that particular subject. This is how we approach the issue. Considering GP as a connection between adaptive agents and their surroundings, from the agent's viewpoint, we shall first show that GP can give us a "subjective" measure of the complexity of the agent's surroundings, and being a model builder, we will then argue that the "subjective" measure can be "objectively" determined by the culture of the society where the agent is situated.

6 Genetic Programming and Human Learning

6.1 Program Complexity of an Agent's Surroundings: A Subjective Measure

We shall exemplify the first assertion based on Chen and Yeh (1997a). They employed GP to discover the underlying law of motion for some simple chaotic time series. They considered the following three chaotic laws of motion:

$$x_{t+1} = 4x_t(1-x_t), \quad x_t \in [0, 1] \forall t \tag{1}$$

$$x_{t+1} = 4x_t^3 - 3x_t, \qquad x_t \in [-1, 1] \ \forall t \tag{2}$$

$$x_{t+1} = 8x_t^4 - 8x_t^2, \qquad x_t \in [-1, 1] \ \forall t \tag{3}$$

These three laws of motion are different in their *algorithmic size*, i.e., the *length* of their symbolic expression. To see this, we rewrite each of the equations above into the corresponding LISP S-expression.

$$(*(4*(x_t(-1\ x_t)))) \tag{4}$$

$$(-(*4(*x_t(*x_t x_t)))(*3x_t)) \tag{5}$$

$$(+(-(*8(*x_t(*x_t(*x_t x_t))))(*8(*x_t x_t)))1) \tag{6}$$

The *length* of a LISP S-expression is determined by counting the number of elements (atoms) in the string that makes up the S-expression from the left-most to the right-most position. From Eqs. 4–6, the lengths of the LISP S-expressions are 7, 11, and 16, respectively. Therefore, in terms of algorithmic complexity, Eq. 1 is the simplest, while Eq. 3 is the most complex. Chen and Yeh then examined how this difference might affect the performance of GP.

By setting the initial value $x_0 = 0.213$, a time series composed of 50 observations is generated for Eqs. 1–3, respectively. Call them time series 1, 2, and 3. These time series serve as the training data for GP. Four experiments were implemented for each series. For each experiment, they let GP run for 1000 generations. For series 1 and 2, GP was able to discover the underlying law of motion in all four experiments. However, the number of generations required for this discovery was different. For series 1, it took 7, 12, 14, and 19 generations, whereas for series 2, it took 29, 37, 37, and 70 generations. As for series 3, GP failed to discover the law of motion in three out of the four simulations, and for the only success the law of motion was discovered at the 151th generation. These experiments show the effect of the length of the LISP S-expression (the algorithmic complexity of the program) on discovery. A simple hypothesis was then proposed in their paper: *The chance of discovering simple (short) programs in a small number of generations is relatively higher than that of discovering long (complex) programs.*

While this hypothesis sounds intuitive, a formal test of it is not that straightforward. One of the technical reasons is that the complexity of a program is not independent of *the terminal set or the function set* initially given by the user. To see this, let us go back to the chaos experiment discussed above. The function set originally employed by Chen and Yeh (1997a) is $\{+, -, \times, \%\}$, and the terminal set is $\{x_t, \mathcal{R}\}$, where $\mathcal{R}$ is the ephemeral random constant.[1] If we add the function "cubic" to the function set, then the minimal description of Eq. 2 is

$$(-(*4(\text{cubic } x_t))(*3x_t)) \tag{7}$$

and the program length is only eight. Alternatively, if we add x_t^3 to the terminal set, then the minimal description becomes

$$(-(*4x_t^3)(*3x_t)) \tag{8}$$

In this case, the program length of Eq. 2 is even shorter and is seven, which is the same as that of Eq. 1. It is therefore likely to discover the hidden law of series 2 as fast as to discover that of series 1. So depending on the user-supplied function set and terminal set, a mathematical function can have different program lengths. Therefore, a refinement of the hypothesis proposed above is given as follows. Let us define the program complexity of a mathematical function *with respect to a given function set and terminal set* as the length of its *minimal description. Then the chance of discovering the mathematical function with the given function set and terminal set in a finite number of generations is an inverse function of its program complexity.*

6.2 Complex or Simple: An Objective Measure Based on Culture

While the omission of relevant functions and terminals can make the description of a hidden law (a mathematical function) extraordinarily complex and hence difficult to learn, this does not suggest that one should start with a sizeable function set and terminal set. Indeed, the inclusion of irrelevant functions or terminals can enlarge the search space, which may make a target more difficult to reach. Therefore, while large function and terminal sets are more likely to reduce the program complexity of a mathematical function and make it easier to find, the larger search space induced may have the net effect of making the chance of discovering it uncertain. Perhaps there is no need for users to worry about whether their choice of function sets and terminal sets is "*optimal*." As we have already mentioned, GP is better regarded as a simulator which simulates human thinking and discovery processes. If humans can make mistakes and learn very slowly in some situations, there is no point in asking GP to be a panacea. As a simulator, what GP can do is to spell out the factors affecting discovery, such as the *program complexity* discussed above. In addition, Szpiro (2000) gave a thorough discussion of a list of factors, some of which were found to be of importance in his two earlier studies (Szpiro 1997a,b). These factors are the size of the func-

[1] See Koza (1992a) for details of the ephemeral random constant.

tion set, the terminal set, the initial population, the mutation rate, and the hold of the best-so-far program.

Let us review the first two factors on this list, namely, the function set and the terminal set. A function in the function set can be regarded as a *subroutine*, and a terminal in the terminal set can be treated as a *concept*. Those subroutines and concepts together represent the *culture* of a society. An advanced society may have accumulated a large number of well-structured subroutines with highly constructive concepts. To model an artificial adaptive agent in this society, one need not start with a primitive function and terminal set. For example, to model an artificial adaptive agent in modern financial markets, one can assume the availability of advanced pricing formulas, and start from there. In their study of option pricing formulas, Chidambaran et al. (2000b) included the Black–Scholes model, as an existing subroutine, into their function set. The formulas generated by GP for the equity options were then adaptations of the Black–Scholes model.

Since sizeable function and terminal sets induces a huge search space for the agent, it is desirable to have a large number of programs in the initial population. The combined effect of a large population size and large function and terminal sets is just a corollary of an advanced civilization: to tackle a problem, citizens of an advanced civilization tend to come up with a lot more ideas (alternatives) than citizens in a less developed society. Nonetheless, after a few generations the population of programs generally becomes very homogeneous, and hence the effective population size is largely reduced. In terms of human behavior, this is not unexpected. If an agent is put down in a stable situation, they will evolve to rely on a few fixed rules, usually called *routines*, without much diversity.

However, a problem technically known as *premature convergence* or *trapping into a local optimum* can be another consequence of the reduction in diversity. While this is also not inconsistent with human behavior, considering that humans are frequently deceived by their eyes and intuitions, it is important to introduce novelties into human cognition. In GP, novelties are introduced into an agent's cognition via the crossover operator and the mutation operator. In most applications in economics and finance, the crossover rate and mutation rate are set as *constants*. While genetic programmers from an engineering background have shown the significance of not setting these as constants, and Szpiro (2000) also made a similar suggestion, how those flexible designs can meaningfully be interpreted as a real human reasoning process is an issue yet to be addressed.

6.3 Features of Ideas-Generation Processes

While by varying the size of the initial population, the terminal set, and the function set one is able to capture artificial agents initially endowed with different civilizations, the capacity of GP to simulate human behavior has to be judged by the *process of ideas* being generated. Note that we are interested not only in the *final outcome* (the discovery), but also in the *process* converging to the outcome. The quotation below may help explain what this criterion means.

> Not only did the genetic programming system manage to "rediscover" Kepler's famous third law, . . . , but also, as the system climbed up the fitness scale, one of its interim solutions corresponds to an earlier *conjecture* by Kepler, published ten years

before the great mathematician finally perfected the equation! (Levy 1992 p. 179; italics added)

Here, not only did GP discover Kepler's third law in its mathematically perfected version, but along the way it also discovered the *conjecture*, i.e., the version before the perfection. Of course, there is no point in expecting GP always to be so precise in replicating a discovery process, since the discovery process itself may be highly random, and hence not unique. Instead, as a first approximation, it would be crucial to see whether the ideas-generation process driven by GP is similar in spirit to the human reasoning process. For this purpose, it would be useful to summarize certain general features of human reasoning.

Generally speaking, the human reasoning process can be characterized as a process *from the simple to the complex*, *from the major to the minor* (*from the heavy to the light*), and *from the linear to the nonlinear*. If the words "simple" and "complex" can be formally defined in terms of *program complexity*, then the ideas-generation process driven by GP is indeed a process from the simple to the complex. It is mainly due to the trees-generation mechanism, which implicitly follows a Possion-like process; therefore, the search process is initially biased toward simple programs. Unless simple programs fail to work, complex programs would find it difficult to become popular in the initial stage. The feature of "*simple-to-complex*" can be found quite extensively in the videotape accompanying the book by Koza (1992a) (Koza and Rice 1992).

Secondly, if the meanings of "major" and "minor" are consistent with the fitness, i.e., the major part corresponds to a much larger fitness than the minor part, then the ideas-generation process driven by GP should present the second feature, *from the major to the minor*. In a practical application, it may get stuck in the major part and be unable to move further, but it is unlikely to be locked in the minor part.[2] While the second feature has not been well acknowledged or studied by genetic programmers, it has motivated a few performance boosters for GP, such as *the two-stage symbolic regression*, *cross breeding*, etc. (Szpiro 2000).

Maybe the most difficult part for GP to satisfy is the third property, "from linear to nonlinear." As mentioned above, GP tends to start with simple functions, and then gradually moves toward complex functions if necessary. Nonetheless, GP is unable to distinguish a simple linear function from a simple nonlinear function. For example, the function ex is as simple as $\beta_1 x$, even though the latter is linear and the former is not. As a result, the chances of having ex or $\beta_1 x$ are basically the same in the initial generation of ideas. However, if this is the case, then given the density of linear functions, the probability of generating a linear function in the initial stage is rather small. In other words, just as it is biased toward simple functions, GP is also biased toward *nonlinear* functions.

The bias toward nonlinearity has created a lot of difficulties in the modeling of human learning and adaptation. For example, regular GP often leads to overly complex nonlinear functions. Such nonlinear functions, comprised of complex combinations of functions and indicators, are often difficult to understand and interpret.

---

[2] For example, using the Rossler attractor with a regular fitness function, Szpiro (2000) showed that GP could "discard" 90% of observations simply because they are *minor*. By doing that, GP could keep track of the remaining 10% even though they were sometimes described as outliers.

In many applications, deciphering the winning programs is an almost impossible task. As a result, it might be futile to use these programs (rules) to understand the processes by which humans behave. The usefulness of the GP method for modeling human learning and adaptation thus seems quite restricted a priori.

One way to fix this problem is to make the terminal set and the function set adaptive also. One may start with a primitive function and terminal set at the beginning of learning, and may enlarge it later when obliged to do so. In this way, we are able to choose a convincing sequence of function sets and terminal sets so that the sequence of rules generated can mimic human learning more closely. For example, if one considers that humans generally start their search from a space of linear functions, not just simple functions, then one can first include only "+" and "×" into their function set. By further restricting the generation rules appropriately, only functions in the form of Eq. 9 will appear, and other complex and nonlinear functions will not show up at this stage.

$$a_0 + \sum_{i}^{n} a_i x_i \tag{9}$$

7 Concluding Remarks

This paper considers an important contribution initiated by John Holland and further extended by John Koza. Based on a biological foundation, an artificial adaptive agent is introduced to economic modeling. This agent is characterized by its capacity for autonomous discovery (autonomous programming ability). This idea has been applied extensively to economics and finance, and we have reviewed a few of those cases. Owing to limitations of space, what has not been covered here is the study of the behavior of a collection of GP-based agents in a market-like environment. Studies in this area can be found in Andrews and Prager (1994), Chen and Yeh (1996b, 1997b, 1999, 2000a, b), and Lensberg (1999). This subject will be covered in the near future.

References

Allen F, Karjalainen R (1999) Using genetic algorithms to find technical trading rules. J Financ Econ 51:245–271
Andrews M, Prager R (1994) Genetic programming for the acquisition of double auction market strategies. In: Kinnear K (ed) Advances in genetic programming, vol 1. MIT Press, Cambridge, pp 355–368
Arifovic J (1991) Learning by genetic algorithms in economic environments. PhD Thesis, University of Chicago
Arifovic J (1994) Genetic algorithms learning and the cobweb model. J Econ Dyn Control 18:3–28
Arifovic J (1995) Genetic algorithms and inflationary economies. J Monetary Econ 36:219–243
Arifovic J (1996) The behavior of the exchange rate in the genetic algorithm and experimental economies. J Polit Econ 104:510–541
Arifovic J, Eaton B (1995) Coordination via genetic learning. Comput Econ 8:181–203
Arthur B (1992) On learning and adaptation in the economy. Sante Fe Institute Economics Research Program, Working Paper 92-07-038

Chen S-H, Yeh C-H (1996a) Genetic programming and the efficient market hypothesis. In: Koza J, Goldberg D, Fogel D, Riolo R (eds) Genetic programming 1996. Proceedings of the 1st Annual Conference. MIT Press, Cambridge, pp 45–53

Chen S-H, Yeh C-H (1996b) Genetic programming learning and the cobweb model. In: Angeline P (ed) Advances in genetic programming, vol. 2. MIT Press, Cambridge, pp 443–466

Chen S-H, Yeh C-H (1997a) Toward a computable approach to the efficient market hypothesis: an application of genetic programming. J Econ Dyn Control 21:1043–1063

Chen S-H, Yeh C-H (1997b) Modeling speculators with genetic programming. In: Angeline P, Reynolds R, McDonnell J, Eberhart R (eds) Evolutionary programming VI. Springer, Berlin, pp 137–147

Chen S-H, Yeh C-H (1999) Modeling the expectations of inflation in the OLG model with genetic programming. Soft Comput 3(2):53–62

Chen S-H, Yeh C-H (2000a) Simulating economic transition processes by genetic programming. Ann Oper Res 97:265–286

Chen S-H, Yeh C-H (2000b) Evolving traders and the business school with genetic programming: a new architecture of the agent-based artificial stock market. J Econ Dyn Control 25:363–393

Chen S-H, Lee W-C, Yeh C-H (1999) Hedging derivative securities with genetic programming. Int J Intell Syst Account Financ Manage 8:237–251

Chidambaran N, Lee C, Trigueros J (2000a) Option pricing via genetic programming. In: Abu-Mostafa Y, LeBaron B, Lo A, Weigend A (eds) Computational finance. Proceedings of the 6th International Conference. MIT Press, Cambridge

Chidambaran N, Triqueros J, Lee C-W (2000b) Option pricing via genetic programming. In: Chen S-H (ed) Evolutionary computation in economics and finance. Physica Verlag, Heidelberg (in press)

Grossman SJ, Stiglitz J (1980) On the impossibility of informationally efficient markets. Am Econ Rev 70:393–408

Holland J (1975) Adaptation in natural and artificial systems. University of Michigan Press, Ann Arbor

Holland J, Miller J (1991) Artificial adaptive agents in economic theory. Am Econ Rev 81:365–370

Holland J, Holyoak K, Nisbett R (1986) Induction: processes of inference, learning and discovery (computational models of cognition and perception). MIT Press, Cambridge

Keber C (1999) Genetically derived approximations for determining the implied volatility. OR Spektrum 21:205–238

Keber C (2000) Option valuation with the genetic programming approach. In: Abu-Mostafa Y, LeBaron B, Lo A, Weigend A (eds) Computational finance. Proceedings of the 6th International Conference. MIT Press, Cambridge

Kiyotaki N, Wright R (1989) On money as a medium of exchange. J Polit Econ 97:927–954

Koza J (1992a) Genetic programming: on the programming of computers by means of natural selection. MIT Press, Cambridge

Koza J (1992b) A genetic approach to econometric modelling. In: Bourgine P, Walliser B (eds) Economics and cognitive science. Pergamon, Oxford pp 57–75

Koza J, Rice J (1992) Genetic programming: the movie. (Videotape accompanying the book *Genetic Programming*: on the programming of computers by means of natural selection, by John Koza.) MIT Press, Cambridge

Lensberg T (1999) Investment behavior under Knightian uncertainty—an evolutionary approach. J Econ Dyn Control 23:1587–1604

Levy S (1992) Artificial life: a report from the frontier where computers meet biology. Vintage, New York

Lucas R (1986) Adaptive behaviour and economic theory. In: Hogarth R, Reder M (eds) Rational choice: the contrast between economics and psychology. University of Chicago Press, Chicago, pp 217–242

Marimon R, McGrattan E, Sargent T (1990) Money as medium of exchange in an economy with artificially intelligent agents. J Econ Dyn Control 14:329–373
Neely C, Weller P, Ditmar R (1997) Is technical analysis in the foreign exchange market profitable? A genetic programming approach. J Financ Quant Anal 32:405–427
Palmer R, Arthur W, Holland J et al. (1994) Artificial economic life: a simple model of a stockmarket. Phys D 75:264–274
Sargent T (1993) Bounded rationality in macroeconomics. Oxford University Press
Szpiro G (1997a) Forecasting chaotic time series with genetic algorithms. Phys Rev E 55:2557–2568
Szpiro G (1997b) A search for hidden relationships: data mining with genetic algorithms. Comput Econ 10:267–277
Szpiro G (2000) Tinkering with genetic algorithms: forecasting and data mining in finance and economics. In: Chen S-H (ed) Evolutionary computation in economics and finance. Physica Verlag, Heidelberg (in press)
Waldrop M (1992) Complexity: the emerging science at the edge of order and chaos. Simon and Schuster, New York

Social Interaction and Coordination Failures in a Simple Model of Technology Adoption

Domenico Delli Gatti* and Mauro Gallegati†

Summary. We analyze how alternative interactive frameworks affect coordination among agents in the equilibrium selection process within a simple model of technology adoption. In particular, we study the dynamics of the coordination efforts of a fixed number of heterogeneous agents whose choices are modeled through genetic algorithms in a setting of bounded rationality and limited information. Differences in the environment are due to the way in which each agent interacts with neighbors in exchanging information. Local stochastic interaction makes coordination on the Nash equilibria easier, and is therefore more efficient than the local deterministic and global stochastic interactive schemes.

Key words. Social interaction, Coordination failures, Technology adoption, Multiple equilibria, Genetic algorithm

1 Introduction

Recent macroeconomic research focuses on multiple equilibria and coordination failures as an explanation of a number of stylized facts, from persistence of nominal and real shocks over the business cycle, to the occurrence of poverty traps in long-run development. While the conditions under which Pareto-ranked multiple equilibria can arise are well known, far less is known on the issues of how, and how fast, agents can succeed in coordinating their actions in order to achieve a Pareto-superior equilibrium when information is limited (see Cooper 1999).

Consider a game that displays multiple equilibria. A coordination problem (Cooper and John 1988) exists when an equilibrium is Pareto-dominated by another one: a universally better outcome can be achieved simply by coordinating the activity of otherwise noncooperative agents. Recent theoretical research (Ellison 1993; Kandori

* Faculty of Economics, Catholic University, Largo A. Gemelli 1, 20123 Milan, Italy
† Faculty of Law, University of Teramo, via Memmingen, Colleparco, 64100 Teramo, Italy

et al. 1993) has shown that the nature—i.e., local or global—of the interactions within a population plays a crucial role in determining how, and how fast, players will achieve coordination (see also Keser et al. 1998).

In this chapter, we analyze how alternative interactive frameworks affect coordination toward equilibria, i.e., the equilibrium selection process, in a simple model of technology adoption. The literature on this issue strongly emphasizes the complementarities in implementing new technologies (Blume 1993). As a consequence, the adoption of a new technology requires a sort of coordination among a large number of agents: only when a critical mass of innovators is reached does the adoption of a new technology become profitable. How does a decentralized economy with limited information develop a coordination mechanism? Our conjecture is: through agents' interactions.

In particular, we study the dynamics of the coordination efforts of a fixed number of heterogeneous agents in different environments with bounded rationality and limited information. We can envisage different environments depending upon the interactive schemes which rule the exchange of information among agents and their neighbours (a similar structure can be found in repeated prisoners' dilemma games: Lindgren 1997; Albin and Foley 1998).

To do so, we build a very simple framework in which boundedly rational agents (spatially located onto a lattice) can choose between two competing technologies. Interaction is the core of the model, as in statistical mechanics. Unlike statistical mechanics, however, interaction itself may change through time. In fact, economic models are populated by agents who continuously make and revise their choices (sell a bond, buy a warrant, etc.), while the subatomic elementary particles interact in obedience to their nature (they cannot decide with whom to interact). While in the standard model the behavior of a part of the system determines aggregate dynamics, we will show that in this model the aggregate determines the individual's behavior.

We will show how fluctuations may be interpreted as coordination failures, and how the technological adoption process generates a self-organized critical system (Foellmer 1974; Durlauf 1997). In particular, gradual changes in the "degree" of interaction can produce discontinuous aggregate dynamics; there is, in fact, a parameter region where criticality arises (when the costs exceed a certain threshold). As Kauffman (1993) notes, the performance of the system is optimized when the process is "at the edge of order and chaos." In particular, when the system is "near" equilibrium, a shock generates fluctuations of a not too relevant magnitude: the production of disorder (or entropy) is very low, and the system regains the previous equilibrium position (in such a case, the standard tools, i.e., stochastic linear difference equations of a very low order, are adequate). As the system moves away from equilibrium, entropy increases, fluctuations become stronger, and new and different forms of equilibrium may emerge, increasing the "transition phase" dynamics (and nonlinear equations become the standard tool: see Prigogine and Stengers 1984; Rosser 1999).

The chapter is organized as follows. In Sect. 2, we present the basic features of the economy under scrutiny. Section 3 is devoted to the behavioural strategies of the agents. In Sect. 4, we present and discuss the results of the computer simulation, and Sect. 5 is a conclusion.

2 Framework

We model an economy with N firms and L workers. The price level is constant and normalized to 1. Each firm has access to a given quantity of labor, $l = L/N$, and can adopt one of two alternative constant-returns technologies, A or B, which use only labor as an input. Owing to an innovation, technology B is more productive than technology A. Since the firm employs a constant amount of labor, output levels are determined solely by the technology chosen. The production functions which embody the two technologies are

$$y_A = al \tag{1}$$

$$y_B = bl \tag{2}$$

where y_i is output obtained using technology i (i = A, B), and a and b are the given and constant returns to labor for the two technologies (b > a).

Labor is remunerated at a constant real wage w. The wage bill or labor cost for each firm is wl. Technology A can be adopted without incurring additional costs. However, in order to adopt the innovative and more productive technology B, the firm must incur extra costs, which we label *innovation costs*. There are two types of innovation cost. The first one is *research and development (R&D) expenditure*, which decreases with the number of firms adopting B (innovators for short). The negative relationship between individual R&D expenditure and the number of innovators is due to a technological spillover: the diffusion of technology B makes its adoption less costly because information about the innovation is shared among innovators. The second innovation cost is a *protection cost* sustained by the firm precisely in order to limit the spillover of technological know-how to other firms; this cost increases with the number of innovators.[1]

According to the above-mentioned hypotheses, the cost functions relative to the two technologies are

$$C_A = wl \tag{3}$$

$$C_B = wl + \frac{R}{n} + cn \tag{4}$$

where R is the given economy-wide R&D expenditure, n is the number of innovators, and c is the unit cost of protection.

The individual profits realized by firms using technologies A and B, respectively, are

$$\pi_A = y_A - wl = l(a - w) \tag{5}$$

$$\pi_B = y_B - wl - \frac{R}{n} - cn = l(b - w) - \frac{R}{n} - cn \tag{6}$$

We assume that $a > w$. Therefore, $b > a > w$.

[1] Protection costs capture the negative effect of the introduction of an innovation on profits. There could be other reasons for such an effect. In a more sophisticated model, for instance, it could be due to saturation of demand for the goods produced by the innovators.

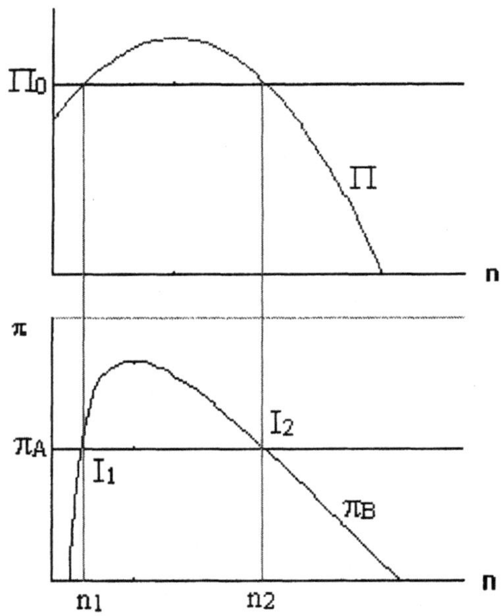

FIG. 1. The *upper panel* shows the plot of aggregate profit Π as a nonmonotonic concave function of the number of innovators. Π_0 is the aggregate profit when no firms adopt the innovation. The *lower panel* shows the plot of individual profits when a firm adopts the new technology (π_B) and when it sticks to the old one (π_A)

π_A is constant, while π_B is a nonmonotonic concave function of the number of innovators. These properties of the profit function in cases where the new technology is adopted are due to innovation costs. When the number of innovators is relatively low (i.e., smaller than $\hat{n} = \arg\max_n \pi_B$), an increase in the number of innovators brings about an increase in π_B: the positive effect on profit of the decrease of average R&D expenditure offsets the negative effect of the increase of protection costs. Symmetrically, when the number of innovators is large (i.e., greater than $\hat{n}$), an increase in the number of innovators brings about a decrease of profits because the positive impact through R&D expenditure is offset by the negative effect due to protection costs.

The new technology is more profitable than the old one, i.e., $\pi_B > \pi_A$, if

$$R + cn^2 < l(b - a)n \tag{7}$$

The firm is uncertain about adopting or not adopting the new technology if

$$cn^2 - l(b - a)n + R = 0 \tag{8}$$

In the lower panel of Fig. 1, the curve represents Eq. 6 as a function of n, while the straight line represents Eq. 5. There are two indifference points (I_1 and I_2) at which the use of technology B yields the same profit as technology A. This happens when the number of innovators is

$$n_1 = \frac{l(b-a)}{2c} - \sqrt{\frac{l^2(b-a)^2}{4c^2} - \frac{R}{c}} \tag{9}$$

or

$$n_2 = \frac{l(b-a)}{2c} + \sqrt{\frac{l^2(b-a)^2}{4c^2} - \frac{R}{c}} \tag{10}$$

Of course, n_1 and n_2 are the solutions of Eq. 8. We assume that $l^2(b-a)^2/4c > R$, so that there are two distinct (positive) roots. Moreover $n_2 \leq N$.

The figure shows that for $n < n_1$ and $n > n_2$, technology A yields higher profits than (and therefore is preferred to) technology B. In these cases, of course, n will decrease. The opposite holds true if $n_1 < n < n_2$: in this case B is preferred to A and n will increase. Therefore, I_1 is unstable, and I_2 is stable. If $n < n_1$, it is clear that the number of innovators goes to zero as time goes by. We therefore count two Nash equilibria characterized by $n_1^* = 0$ (no innovation equilibrium thereafter) and $n_2^* = n_2$, respectively.

These Nash equilibria can be Pareto-ranked. In fact, if we assume that all the income received by households is spent on the consumption of goods, and utility increases with consumption, aggregate income can be conceived as a rough measure of social welfare. Aggregate income is the sum of the income produced by A-firms ($Y_A = (N-n)y_A = (N-n)al = aL - nal$) and the income produced by B-firms ($Y_B = nbl$), i.e.,

$$Y = aL + l(b-a)n$$

Therefore aggregate income increases with the number of innovators. As a consequence, aggregate income (and social welfare) in the no-innovation equilibrium is $Y_1^* = aL$, while in the equilibrium with innovation it is $Y_2^* = aL + l(b-a)n_2^*$. It is clear that the equilibrium with innovation is Pareto-superior.

Recalling that $n_2 \leq N$, we conclude that in the equilibrium with innovation, not all firms necessarily adopt the new technology. Since income increases with the number of innovators, the maximization of income and social welfare requires the adoption of the new technology by all firms. Therefore, social welfare in the Pareto-superior Nash equilibrium is not necessarily the highest attainable in principle. In other words, even if the firms settle in Pareto-superior equilibrium, there could still be social inefficiency due to the survival of firms adopting the old technology.

This conclusion is due to the nonmonotonicity and concavity of the profit function of the innovator, which in turn can be traced back to the protection costs, as stated above. Without protection costs, the curve representing the profit of the innovator would be upward sloping, there would be only one intersection (under appropriate conditions) with the straight line representing the profit of the noninnovator, and therefore only one critical mass of innovators: for each initial condition (number of innovators) smaller than this threshold, the economy would converge to the no-innovation equilibrium, while for each initial condition greater than the threshold, the innovation would spread to all the firms in the economy. In other words, there would be two polar Nash equilibria in which all the firms either adopt the old technology or switch to the new one. In the present model there are two critical masses of innovators, and the equilibrium with innovation is characterized by the survival of some noninnovators: when the number of innovators is high, protection costs can become prohibitive and induce some firms to stick to the old technology.

Aggregate profits for noninnovators and innovators, respectively, are

$$\Pi_A = (N-n)\pi_A = (L-nl)(a-w) \tag{11}$$

$$\Pi_B = n\pi_B = nl(b-w) - R - cn^2 \tag{12}$$

Aggregate economy-wide profit can be obtained by summing Eqs. 11 and 12:

$$\Pi = (N-n)\pi_A + n\pi_B$$
$$= L(a-w) + nl(b-a) - R - cn^2 \quad (13)$$

where R is zero if there are no innovators. It is interesting to note that aggregate profit is a nonmonotonic concave function of the number of innovators. These properties of the aggregate profit function are due to innovation costs.

If all agents adopted technology A (which happens in the no-innovation equilibrium), aggregate profits would be

$$\Pi_0 = N\pi_A = L(a-w)$$

Aggregate profits when there are both innovators and noninnovators are greater than aggregate profits in the no-innovation equilibrium if

$$nl(b-a) - R - cn^2 > 0 \quad (14)$$

It can then be concluded that Eq. 14 is the same as Eq. 7. Therefore, for $n < n_1$ and $n > n_2$, the coexistence of technologies A and B yields lower aggregate profits than the widespread adoption of the old technology. The opposite holds true if $n_1 < n < n_2$.

In the upper panel of Fig. 1 we depict the aggregate profit curve and the straight line representing Π_0.

Individual optimization choices are incompatible with social efficiency because of production externalities (different optimal levels of individual profit and aggregate income; Arthur 1988). In other words, one may say that the externalities are the "social cost" of the extra income produced by B.

In the next section, we formalize the behavioral strategies of agents on the basis of the information to which they have access. The agents' behavior has been generated by computer simulations, while the agents' mental models have been modelled by means of a genetic algorithm (Holland 1975).

3 Behavioral Strategies

In a limited information/bounded rationality setting, an agent—say the i-th firm—formulates a strategy (which consists of adopting or not adopting the new technology) according to the information available. This information consists of

— the agent's own profit gained in the previous period π_{it-1}, which is private information;
— the average profit gained in the previous period by a certain number m of agents whom the first agent has "spied" on (π_{jt-1}; $j = 1, 2, \ldots, m$); this information is "limited," i.e., by interacting with the neighborhood the agent may learn the average profit—say $\bar{\pi}_{jt-1}$—but not the standard deviation of the profits of neighbors (the interaction with neighbors can be local or global, deterministic or stochastic, see Sect. 4 for details);
— aggregate profit in the previous period Π_{t-1}, which is public information, i.e., known to everyone in the economy.

Analogously to statistical mechanics (Brock and Durlauf 1995; Durlauf 1997), we must attach a weight to each of these pieces of information and construct a choice function which takes account of each of them, suitably weighted. Agents, however, are not molecules with their own specific weight: they can make choices and continuously revise their plans through an interactive "trial and error" process. Therefore, unlike statistical mechanics, weights evolve through time.

Since the choice involves the evaluation of only two alternatives, we can construct the choice function probabilistically:

$$F_{it} = F_{it}(\pi_{it-1}, \overline{\pi}_{jt-1}, \Pi_{t-1}, \alpha_{it}, \beta_{it}, \gamma_{it})$$

where $0 \le F_{it} \le 1$. α_{it}, β_{it}, and γ_{it} are the weights associated with each variable. Of course, $\alpha_{it} + \beta_{it} + \gamma_{it} = 1$.

If we denote technology B numerically by 1 (one) and technology A by 0 (zero), the agent will choose B if $F_{it} > 1/2$ and A otherwise. Following the statistical mechanics literature, we express the function as a weighted sum:

$$F_{it} = \alpha_{it} f(\pi_{it-1}) + \beta_{it} g(\overline{\pi}_{jt-1}) + \gamma_{it} h(\Pi_{t-1})$$

The strategy actually pursued by the agent depends upon available information. The weights evolve according to a genetic algorithm which allows selection among different values attributable to weights. In essence, when processed by the algorithm, the weight undergoes a change in value (partly deterministic and partly stochastic) which is positively correlated with the profit that the agent has obtained in the previous period. For example,

$$\alpha_{it+1} = \Phi(\pi_{it}); \qquad \Phi' > 0$$

where the functional link is represented by the genetic algorithm.

As regards the specific form of *f*, *g*, and *h*, we have used a simple system of rules which we present and discuss in the following three subsections.

3.1 The Functional Form of f

As far as *f* is concerned, the following rules apply:

— if an agent has adopted A in the past, he/she will continue to do so in the future because he/she does not know whether B is better or worse than A;
— if an agent has adopted B in the past, and the profit has proved to be smaller than that achievable by adopting A, he/she will switch to A;
— if an agent has adopted B in the past, and the profit has proved to be the same as that achievable by adopting A, he/she will continue to use A (*inertia principle*).

Therefore *in principle* the agent would adopt B if B were more profitable than A:

$$f = \begin{cases} 1 & \text{if } \pi_{it-1} > \pi_A \\ 0 & \text{otherwise} \end{cases}$$

In running the simulations, the *f*-function is slightly more complicated: *f* assumes value 1 with a probability equal to the ratio of the number of cases in which $\pi_{it-1} > \pi_A$ and the total number of cases.

3.2 The Functional Form of g

An agent knows the average profit of a certain number of other agents (neighbors). Although some of them gain a profit which is different from $\pi_A = l(a - w)$, the differences between their levels of profit and π_A must necessarily have the same sign, given that there are only two observable behaviors. In other words, the average value of the neighbors' profits is certainly representative of the sign of the difference between the contingent profits of B and A, and here the sign is the only relevant piece of information.

Consequently, the agents will not alter their strategy if their pay-off is greater than, or the same as the average of "spied on" neighbors. If the pay-off is less, the agent will change the strategy. Using s to denote the technology adopted in $t - 1$, we can write

$$g = \begin{Bmatrix} s & \text{if } \pi_{it-1} \geq \overline{\pi}_{jt-1} \\ 1-s & \text{otherwise} \end{Bmatrix}$$

3.3 The Functional Form of h

If the level of aggregate profit is lower than that obtained if all firms use A ($\Pi_0 = (a - w)L$), there are either "too few" or "too many" firms which use B. Because no agent knows the exact number of firms which use B, they cannot know whether the system lies below the second indifference point or the first one. If they opts for B, and the behavior of the others remains constant, the following situations may arise:

— there are too many users of B, so that one more user would further reduce income, and with it profits;
— there are too few users of B, so that one more user pushes up income and profits, but the latter can at most be equal to the profits deriving from A.

Therefore the strategy "adopt B" has an expected value that is lower than the (certain) value of A: in this situation it is convenient to use A.

If instead the profit level is higher than Π_0, B is the dominant strategy. In fact, when the system lies on the downward sloping branch of the individual profit curve, the addition of one user of B may at most make it as economically convenient as A: the expected value of B is higher than the certain value of A.

Finally, if the profit level is exactly equal to Π_0, this means either that all firms are using A (and therefore switching to B would yield negative profits because of the innovation costs), or that the system is at one of the indifference points. In this case, the agent does not have enough information to identify the exact point, and we keep things simple by assuming that they continue to use the same technology employed in the previous period (*inertia principle*). Using s to denote the technology adopted in $t - 1$, we have

$$h = \begin{Bmatrix} 0 & \text{if} & \Pi_{t-1} < \Pi_0 \\ 1 & \text{if} & \Pi_{t-1} > \Pi_0 \\ s & \text{if} & \Pi_{t-1} = \Pi_0 \end{Bmatrix}$$

TABLE 1. Mean and standard deviation (in parentheses) of the number of innovators during four simulations with different types of interaction

Periods	DLI	SGI	SLI	NI
1–300	69.96 (16.61)	69.51 (27.46)	74.1 (12.47)	66.42 (10.61)
1–100	75.9 (12.28)	68.63 (29.56)	74.12 (13.01)	66.04 (11.84)
101–200	67.56 (17.99)	67.59 (28.84)	74.21 (13.47)	65.74 (10.75)
201–300	66.42 (17.47)	72.3 (24.47)	73.97 (10.92)	62.49 (9.07)

4 Computer Simulations

The simulations were conducted assuming that agents are uniformly arranged on a plane, with the same field of territorial domination as *cellular automata*. The system is represented as a grid in which each cell is an agent, and the color indicates the technology adopted. The cells in the marginal rows or columns ideally border on those located on the opposite margin; these conditions at the extremes make the plane topologically identical with the torus.

Four simulations were carried out:

— *deterministic local interaction* (DLI), where each agent can observe the behavior of the neighbors on the basis of the king's move;[2]
— *stochastic local interaction* (SLI), where each agent can observe the behavior of half of the agents randomly selected among the agents close by (arranged accorded to the king's and knight's moves);[3]
— *stochastic global interaction* (SGI), where each agent can observe a constant number of agents chosen at random from all of the players;
— *no interaction* (NI), where each agent makes choices on the basis of private information (i.e., profits) and public information (aggregate output) without directly interacting with other agents.

The parameter values used in the simulations in the following subsections ($a = 3$; $b = 5$; $c = 0.5$; $w = 2$; $N = 100$; $L = 2000$; $R = 200$) determine the indifference values of n in the use of the two technologies, which are $n_1 = 5$ and $n_2 = 74$, respectively. Initial conditions have been set at $n_0 = 44$ and $n_0' = 2$.

The main results are as follows. Depending on the initial conditions, the system converges to $n_1^* = 0$ or $n_2^* = 74$ (sporadic and temporary convergence to $n = 6$ has also been detected). We may therefore state that the N-player game ensures convergence towards a Nash equilibrium. This result confirms the notion that Nash equilibrium is a powerful predictor of behavior even for N-person games, for which one could reasonably expect randomization of strategies.

Interaction improves coordination (Table 1); in particular, SLI is more efficient than DLI and SGI, while NI is relatively inefficient.

[2]The king can move to all the cells contiguous to the one that he is occupying, horizontally, vertically, and diagonally; in total, he "observes" eight cells.
[3]The knight controls the eight cells closest to it which not do not lie in the same row, column, or diagonal.

We may therefore say that interaction speeds up convergence towards Nash equilibria. The average speed of convergence decreases steadily moving from stochastic global interaction to deterministic local interaction, with the figure for stochastic local interaction being roughly half of that for no interaction. This result can be explained in terms of differences in learning about the strategy profiles of other agents. Owing to repetitions of the baseline one-shot game, subjects are allowed to adapt their beliefs on the equilibrium strategies of their opponents.

Simple nonparametric statistical analysis shows that interaction slightly eases coordination on the payoff-dominant equilibrium. This found support when we applied the two-sided *U*-test to find out whether agents interacting locally or globally tend to choose B more often than those not interacting: the hypothesis of no difference has been rejected at the 1% level. Heuristically, we noted that for agents using global interaction, convergence to the payoff-dominant equilibrium happened in 72% of cases, while the corresponding figure for agents interacting locally was 44%. At odds with experimental evidence, however, both the Nash equilibria, i.e., the risk-dominant and the payoff-dominant ones, were observed with significant frequency.

The dispersion between the share of adopters and aggregate profits is limited to the descending curve of the parabola for NI and SLI alone (with the latter being a little bit larger than the former, which never exceeds $n = 80$), while both DLI and SGI explore both sides of the curve (SLI in particular, examines all the points superior to the first unstable Nash equilibrium). Moreover, the learning process is almost nil for NI agents, while it is positive for the interacting agents.

The convergence towards equilibrium is asymmetric. The expansive phase lasts (with two periods with fluctuating paths, if interaction is taken into account) until the Nash equilibrium point is reached. When n becomes greater than n_2, a massive technological reswitching happens, and the convergence process toward equilibrium restarts. More generally, the dynamic behavior (either in the spatial structure or in time series) is sensitive to the initial condition, and chaotic and complex behavior (in the sense of Day 1994), as well as transition phases, have been detected.

For early stages of the game, i.e., the ones in which learning is likely to matter more, standard deviations for cases in which agents are allowed to communicate with other subjects are substantially higher than those for cases with no interaction. Higher values of the standard deviation during the initial stages of the game can be due to a higher degree of random experimentation or mutation (Kandori et al. 1993) by subjects. According to our conjecture, subjects are more prone to "risk" playing different strategies when they can exchange information with other players. For the NI case, the speed of convergence was relatively low, and a path-dependency phenomenon has been detected.

In the following subsections we describe some computer simulations for fluctuations around n_2.

4.1 Simulations with DLI

The dynamic behavior of the number of innovators in this case is very irregular as far as amplitude is concerned (Fig. 2), and is characterized by transition phases: it exhibits periods in which the system is in the Nash equilibria, or fluctuates above or below it. These fluctuations last for two periods (because agents shift between the two

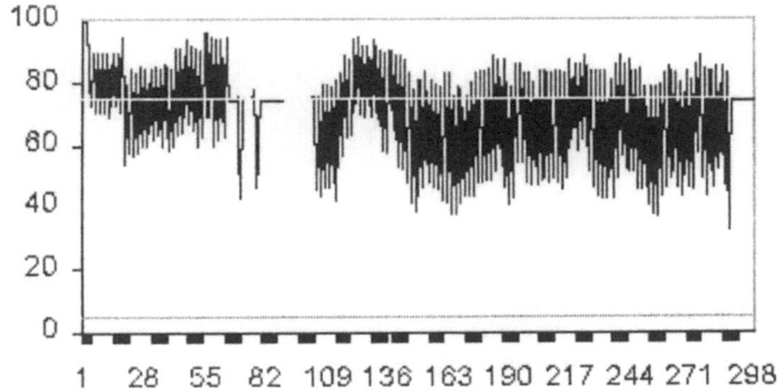

FIG. 2. Time series of the number of innovators when social interaction is of the deterministic local type

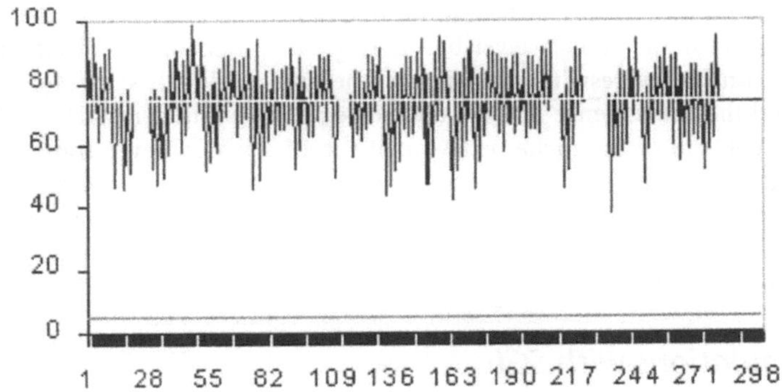

FIG. 3. Time series of the number of innovators when social interaction is of the stochastic local type

technologies), while the dynamics are quite regular in some periods and niche behavior (and therefore hysteresis) emerges. On average it stands 6% below the Nash equilibrium, with a standard deviation around 16%. As far as the interaction parameters are concerned, we note that private information gains momentum at the expense of local information because the latter generates path-dependency, which can be nonoptimal. It is worth noting that the change is more evident just before a change in the transition phase.

4.2 Simulations with SLI

Figure 3 shows the result of a simulation characterized by SLI. This interaction is the most efficient one: its average level stands less than 1% below the Nash equilibrium, with a standard deviation of 12%. The phase transition phenomenon recurrs frequently, and one-dimensional chaotic motion has also been detected by the BDS test. This phenomenon has a frequency of period two, while the amplitude is very

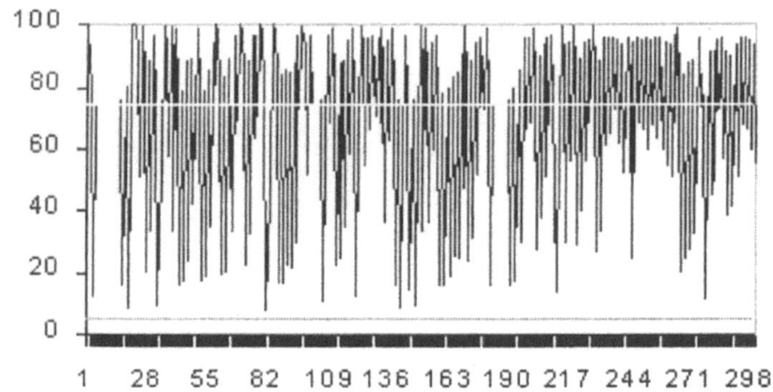

FIG. 4. Time series of the number of innovators when social interaction is of the stochastic global type

variable. As far as the convergence toward equilibrium is concerned, the expansion phase lasts until the trough level becomes greater than the Nash equilibrium. A massive technological reswitching happens (the lattice suddenly shows full–empty cell states), and the convergence process towards equilibrium may restart. The local stochasticity gets rid of the niche behavior of the DLI, which never traps the system: niche behavior is then easily overcome by stochastic interaction. On the other hand, this rewards the evolution of private and local information, whose parameters rise at the expense of the public information element. The dynamics resemble those of the DLI, but are much more locally variable.

4.3 Simulations with SGI

With the chance of meeting all the agents in the torus, the explorative possibilities reach a maximum value. This is reflected in the standard deviation of the convergence, which on average is around 27% (more than twice that of DLI and SLI, and three times that of the NI case), and the amplitude of the fluctuations is the highest of all the four interaction cases (Fig. 4).

Moreover, all the parameters computed by means of the genetic algorithm remain fairly stable. The transition phase phenomenon is also present, even if far less stable than in the previous cases because of the global stochastic element. In the lattice, niche behavior emerges which is similar, but less regular, than that for DLI.

4.4 Simulations with NI

In this simulation (Fig. 5), an agent's behavior is driven by both public and private information. No spatial interaction is allowed. The dynamic pattern toward equilibrium is slow (on average 6–8 times slower than in the SLI case, and after 300 runs, it is still 10% below the Nash equilibrium). As time passes, the periodicity of the fluctuations increases, while still remaining generally below the Nash equilibrium level. This conservative behavior can be explained by risk-aversion, which is exacerbated when the context of no-interaction prevents the spread of additional information.

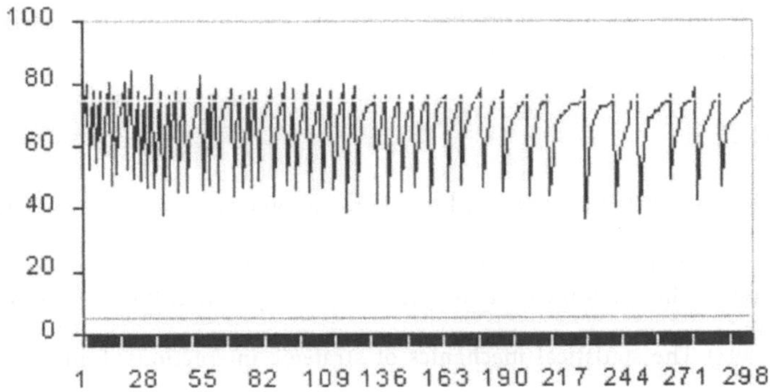

FIG. 5. Time series of the number of innovators when there is no interaction

The aggregate dynamics strictly conform to this rule: the adoption of a new technology slowly spreads through the agents until the Nash equilibrium is reached, a further increase in technology adoption becomes unprofitable, and the cycle turns into a sensitive technological reswitching. Fluctuations are irregular for frequency and amplitude, even if they tend to decrease in frequency with time. Moreover, their standard deviation is relatively low (around 10%) and moderately decreasing through time. In fact, as far as the evolution of the parameters is concerned, agents progressively rely on private information so that niche behavior emerges and innovations progressively slow down. The dynamics replicate this behavior with no regularity, and the changing periodicity mentioned above.

5 Concluding Remarks

We have presented a very simple model of the adoption of technology with multiple Pareto-ranked Nash equilibria in order to study how alternative informational structures and social interaction schemes affect the process of equilibrium selection in a coordination game. Social interaction in our context essentially means information transmission among sets of neighboring players.

We ran simulations with three types of social interations: *deterministic local interaction* (DLI), *stochastic local interaction* (SLI), and *stochastic global interaction* (SGI), and compared the results with the benchmark case of *no interaction* (NI).

Our results confirm the notion that Nash equilibrium is a powerful predictor of behavior even for *N*-person games. Interaction improves coordination; in particular, SLI is more efficient than DLI and SGI. It allows convergence to occur at a very high speed. However, we need more evidence from simulations in different frameworks, or experiments with other games, to see how robust these results are.

Acknowledgments. We are indebted to Alessandra Iacobucci and Antonio Palestrini for in-depth discussions of the issues at stake in modeling social interactions. We also thank participants at the JAFEE meeting in Tokyo and at a seminar held at the University of Nice (in particular R. Arena) for their penetrating comments and criticisms. Michele M. Cruciano has been an excellent research assistant, and moreover he ran all the computer simulations.

References

Albin P, Foley D (1998) Barriers and bounds to rationality. Princeton University Press, Princeton
Arthur WB (1988) Self-reinforcing mechanisms in economics. In: Anderson P, Arrow K, Pines D (eds) The economy as a complex evolving system. Addison-Wesley, Redwood City, pp 9–31
Blume L (1993) The statistical mechanics of strategic interactions. Games Econ Behav 5:387–426
Brockw, Durlaufs (1995) Discrete choice with social interaction. I. Theory. University of Wisconsin, SSRI, wp9521
Cooper R (1999) Complementarities and macroeconomics. MIT Press, Cambridge
Cooper R, John A (1988) Coordinating coordination failures in Keynesian models. Q J Econ 103:441–463
Day R (1994) Complex economic dynamics. MIT Press, Cambridge
Durlauf S (1997) Statistical mechanics approaches to socioeconomic behavior. In: Brian A, Durlauf S, Lane D (eds) The economy as a complex evolving system, vol II. Addison-Wesley, Reading, pp 81–104
Ellison G (1993) Learning, local interaction, and coordination. Econometrica 61:1047–1071
Foellmer H (1974) Random economies with many interacting agents. J Math Econ 1:51–62
Holland JH (1975) Adaptation in natural and artificial systems. University of Michigan Press, Ann Arbor
Kandori M, Mailath GC, Rob R (1993) Learning, mutations, and long run equilibria in games. Econometrica 61:29–56
Kauffman S (1993) The origins of order. Oxford University Press, New York
Keser C, Ehrhart KM, Berninghaus SK (1998) Coordination and local interaction: experimental evidence. Econ Lett 58:269–275
Lindgren K (1997) Evolutionary dynamics in game-theoretic models. In: Brian A, Durlauf S, Lane D (eds) The economy as a complex evolving system, vol II. Addison-Wesley, Reading, pp 337–367
Prigogine N, Stengers I (1984) Order out of chaos. Bantam, New York
Rosser JB (1999) On the complexity of complex economic dynamics. J Econ Perspect 13(4):169–192

Effects of Competitive Metaplayers in the Distributed Social Dilemma

TOMOHISA YAMASHITA* AND AZUMA OHUCHI*

Summary. This paper deals with the extended social dilemma model "distributed social dilemma," which represents a situation in which players can select the game they play and then decide whether to cooperate or defect. In order for mutual cooperative behavior to emerge, one decentralized organization, a "metaplayer," is assigned to one social dilemma game, and each metaplayer can manage a social dilemma game by imposing light punishments. By comparing various cases, we were able to determine which interactions among metaplayers are most effective for the promotion of cooperative behavior. To observe the behavior of player and metaplayer, we use agent-based simulation in which each agent is regarded as player or metaplayer.

Key words. Social dilemma, Game theory, Agent-based simulation

1 Introduction

After the publication of the work of Axelrod (1984), many researchers have become interested in the concept of mutually cooperative behavior among nonaltruistic agents. The iterated prisoner's dilemma (Axelrod 1984, 1997) and the social dilemma (Axelrod 1986; Dawes 1981; Umino 1989) have been used to represent the interaction among nonaltruistic agents. It is more difficult to establish mutual cooperation in a social dilemma situation than in the iterated prisoner's dilemma (Yao and Darwen 1994). In a social dilemma game, a situation in which players select game partners or the game to be played does not arise. Individual players have no control over which opponents they play. Players are always matched with the same game partners and play the same game. However, in the real world, players are not always in a situation where they have no alternative but to play their assigned social dilemma game. Social

* Graduate School of Engineering, Hokkaido University, Kita 13 Nishi 8, Kita-ku, Sapporo 060-8628, Japan

interactions are often characterized by the preferential choice of the game itself (Ashlock et al. 1996; Tesfatsion 1998). Therefore, in order to represent such a situation, we propose the "distributed social dilemma," in which players select one social dilemma game and then decide whether to cooperate or defect (Yamashita et al. 2000).

In previous studies on social dilemmas, individuals have had the right to punish free riders, or to banish them from the group; that is, individuals have had the right to sanction free riders (Ashlock et al. 1996; Schüssler 1989; Tesfatsion 1998; Yamagishi and Hayashi 1996). The sanctions by individuals are based on mutual surveillance and severe punishment (Axelrod 1986). In a small-scale society, such sanctions are possible, but in a large-scale society, a centralized organization generally enforces penal regulations. While many studies have dealt with sanctions imposed by individuals, few studies have dealt with sanctions imposed by a centralized organization (e.g., central government). The reason for this is that collective cooperation is thought to be easy to establish if an altruistic centralized organization, that is, a maximizer of public interests, enforces severe punishment. However, in reality, severe punishment and an altruistic centralized organization are difficult to introduce into a distributed social dilemma situation. Therefore, we consider light punishment enforced by a nonaltruistic decentralized organization, that is, a maximizer of its own profit, as one of the solutions to a distributed social dilemma. In a distributed social dilemma, we assume that one decentralized organization is assigned to one social dilemma game. Furthermore, each decentralized organization can manage a social dilemma game by imposing light punishment, but cannot manage a centralized distributed social dilemma (e.g., local government).

Here, a decentralized organization is defined as a "metaplayer" because a decentralized organization has the direct power to change the payoff function of individuals as "players." In this chapter, we show which interactions among decentralized organizations (i.e., the number of metaplayers and competition or cooperation) are most effective in the promotion of cooperative behavior. The effects of metaplayers (for an increase in the number, the average payoff, and the amount of payoff of cooperating players) depend on the tendency of players to cooperate originally. For example, a certain interaction among metaplayers is more effective on an altruistic player set than on a nonaltruistic player set in promoting cooperative behavior. In order to evaluate the effects of metaplayers, we introduce the concept of "preference of players to cooperate," which specifies the various tendencies of players to cooperate. By comparing various cases of interaction among metaplayers (e.g., metaplayers are competitive or cooperative, or a situation in which metaplayers exist or do not exist), we can determine how the interaction of metaplayers affects the promotion of cooperative behavior. In this chapter, we formulate a distributed social dilemma with metaplayers using game theory. Furthermore, we consider a distributed social dilemma not as a one-shot game, but as an iterated game, because social dilemmas generally depend more on decision-making over a long period than over a short period. For evaluation and comparison, we use an agent-based simulation in which each agent is regarded as a player or metaplayer (Epstein and Axtell 1996; Namatame 1999).

2 Distributed Social Dilemma with Metaplayers

2.1 Competition among Metaplayers

A metaplayer imposes a penalty on defecting players as a punishment, and levies a tax on all players as a penalty cost in the game to which each metaplayer has been assigned. The payoff for the metaplayer is the amount of the tax and the penalty. The purpose of the metaplayer is not to maximize the payoff for the players, but to maximize their own payoff by adjusting both tax and penalty. In a distributed social dilemma, metaplayers cannot raise a tax or impose a penalty selfishly, because players can select the game. The reason for this is that metaplayers obtain no payoff if no player selects the game to which a selfish metaplayer has been assigned. As a result, metaplayers face the dilemma of whether to raise or reduce a tax, and whether or not to impose a penalty. Therefore, competition among metaplayers may prevent them from behaving selfishly, even if the purpose of a metaplayer is to maximize their own payoff.

2.2 Basic Structures

The basic structures of the distributed social dilemma with metaplayers are as given below.

$$\text{Player set } N = \{1, \ldots, i, \ldots, n\} \tag{1}$$

Strategy set of player $\Pi = \{C_j, D_j | j = 1, \ldots, m\}$

$$C_j\text{, cooperate in game } j; \qquad D_j\text{, defect in game } j \tag{2}$$

Payoff function f of player in game j

$$\begin{aligned} &|C_j|\text{, number of players selecting } C_j \\ &|D_j|\text{, number of players selecting } D_j \end{aligned} \tag{3}$$

	$\frac{B}{D} \leq m\frac{\lvert C_j\rvert}{\lvert D_j\rvert+\lvert C_j\rvert}$	$\frac{B}{D} > m\frac{\lvert C_j\rvert}{\lvert D_j\rvert+\lvert C_j\rvert}$
C_j	$\alpha - m\frac{\lvert D_j\rvert}{\lvert D_j\rvert+\lvert C_j\rvert}D - B - \text{Tax}_j$	$\alpha - mD - \text{Tax}_j - \text{Pen}_j$
D_j	$\alpha - m\frac{\lvert D_j\rvert}{\lvert D_j\rvert+\lvert C_j\rvert}D - \text{Tax}_j - \text{Pen}_j$	$\alpha - mD - \text{Tax}_j - \text{Pen}_j$

α, B, and D, coefficients of payoff functions of players

$$\text{Metaplayer set, } M = \{1, \ldots, j, \ldots, m\} \tag{4}$$

Strategy set of metaplayer, Ω_{Tax}, Ω_{Pen}

$$\begin{aligned} &\Omega_{\text{Tax}} = \{\text{tax}_1, \ldots, \text{tax}_l\} \qquad \Omega_{\text{Pen}} = \{\text{pen}_1, \ldots, \text{pen}_l\} \\ &0 \leq \text{tax}_k \leq \text{tax}_{\max} \ (\text{tax}_{\max}\text{, maximum tax}) \\ &0 \leq \text{pen}_k \leq \text{pen}_{\max} \ (\text{pen}_{\max}\text{, maximum penalty}) \end{aligned} \tag{5}$$

Payoff function of metaplayer j, $g_j(\text{Tax}_j, \text{Pen}_j)$

$$g_j(\text{Tax}_j, \text{Pen}_j) = (|C_j| + |D_j|)\text{Tax}_j + |D_j|\text{Pen}_j$$
$$\text{Pen}_j = \{\text{pen}_k | \text{pen}_k \in \Omega_{\text{Pen}}\} \qquad \text{Tax}_j = \{\text{tax}_k | \text{tax}_k \in \Omega_{\text{Tax}}\} \tag{6}$$

In our simulation, we assume a light penalty; that is, the penalty does not directly change the dilemma structure of the payoff function of the players. A defecting strategy always dominates a cooperative strategy even if a penalty is imposed on the payoff function of the players. Consequently, the maximum penalty satisfies the following inequality:

$$0 \le \text{pen}_{\max} \le B \tag{7}$$

In order to construct a simple model, assume that the payoff for player is always more than 0. In the case in which the minimum payoff for players is 0 and the maximum penalty satisfies inequality (7), the maximum tax satisfies the following inequality:

$$0 \le \text{tax}_{\max} \le \alpha - nD - \text{pen}_{\max} \tag{8}$$

3 Decision-Making Mechanism

3.1 Expected Payoff

Individual players and individual metaplayers make decisions based on the "expected payoffs" for each strategy (Ashlock et al. 1996; Tesfatsion 1998). The expected payoffs of players are influenced by the preferences of players to cooperate (Uno and Namatame 1998). When a player or a metaplayer has received a payoff from the game, they can update the expected payoff to reflect the new payoff which will be received. The expected payoff is defined as follows. Suppose that a player (or a metaplayer) v selects a strategy s, the number of players v selecting strategy s is $T_{v,s}$, the t-th payoff in a series of payoffs acquired by strategy s ($X_{v,s} = (x^1_{v,s}, \ldots, x^{T_{v,s}}_{v,s})$ is $x^t_{v,s}$, and the discount factor is δ, then the expected payoff for player v from strategy s is represented as

$$\text{expected payoff}_{v,s} = \sum_{t=1}^{T_{v,s}} \delta^t x^t_{v,s} \tag{9}$$

3.2 Preference of Player

As the preference of players to cooperate is great, they are more likely to cooperate even if they acquire lower payoffs. The preference of player i for strategy s is represented as $p_{i,s}$ and influences expected payoff$_{i,s}$ as follows:

$$\text{expected payoff}_{i,s} = \sum_{t=1}^{T_{i,s}} \delta^t (x_{i,s} + p_{i,s})^t \tag{10}$$

Here, the preference $p_{i,s}$ of player i for strategy s is represented by the equation

$$p_{i,s} = \begin{cases} p(i) & \text{if } s \in C_j \\ 0 & \text{otherwise} \end{cases} \tag{11}$$

The preference for each cooperative strategy is equal. The preference function $p(i)$ is arrived at by using Eq. 12 based on a normal distribution (Uno and Namatame 1998). Equation 12 is the relation between the preference of player p and the number of players $n(p)$.

$$n(p)=\frac{A}{\sqrt{2\pi}}\exp\left[-\frac{1}{2}\left(\frac{p-\mu}{\sigma}\right)^2\right] \tag{12}$$

3.3 Strategy Selection

In our simulation, the rationality of players and metaplayers is the selection of the maximum expected payoff. Player i selects strategy s^* in which expected payoff$_{i,s}^*$ is largest of all the expected payoffs of player i.

$$s^* = \arg\max_{s\in\Pi_i} \sum_{t=1}^{T_{i,s}} \delta^t (x_{i,s}+p_{i,s})^t \tag{13}$$

Similarly, metaplayer j selects strategy s^* in which the expected payoff$_{j,s}^*$ is the largest of all expected payoffs.

$$s^* = \arg\max_{s\in\Omega} \sum_{t=1}^{T_{j,s}} \delta^t x_{j,s}^t \tag{14}$$

4 Experimental Setup

Metaplayers engage in two types of competition. One is competition against other metaplayers. The other is competition against games without metaplayers. If there are both games with metaplayers and games without metaplayers, metaplayers compete against not only other metaplayers, but also games without metaplayers. The reason is that players are more likely to cooperate in a game without a metaplayer as the preference of players to cooperate is greater because there is no tax in such a game.

Under some conditions of competition between metaplayers, we observe how the behavior of players changes as the preference of players to cooperate is greater. Therefore we increase coefficient μ of preference function $p(i)$ one at a time from −40 to 180. According to the range of coefficient μ, if the preference of a player to cooperate is over 160, this player loses the incentive to free ride (defect). If the preference is under −40, this player loses the incentive to cooperate.

In this study, we performed eight kinds of simulation (cases 1–8). The conditions of the eight simulations and the main parameters are shown in Tables 1 and 2.

5 Simulation and Discussion

We evaluated the effects of metaplayers on (1) the number of cooperating players, (2) the average payoff of cooperating players, and (3) the amount of payoff of cooperating players. The results shown in Figs. 1–4 are based on the average for the last 2000 iterations (iterations 28000–30000) of 10 trials.

TABLE 1. Number of metaplayers assigned to the game and the condition of competition among metaplayers in cases 1–8

	Case	1	2	3	4	5	6	7	8
Metaplayer	|M|	0	1	2	2	3	3	4	4
	Game 1	off	on	on	on	on	on	on	on
	Game 2	off	off	on	on	on	on	on	on
	Game 3	off	off	off	off	on	on	on	on
	Game 4	off	off	off	off	off	off	on	on
	Competition	off	off	on	off	on	off	on	off

In this table, "on" means there is a metaplayer, and "off" means there is no metaplayer. If competition is "on," each metaplayer aims to maximize his or her own payoff, and if competition is "off," each metaplayer aims to maximize the amount of payoff for metaplayers. For example, in case 5, metaplayers are assigned to games 1, 2, and 3, and they each aim at maximizing their own payoffs, while there is no metaplayer in game 4, i.e., $Tax_4 = 0$ and $Pen_4 = 0$

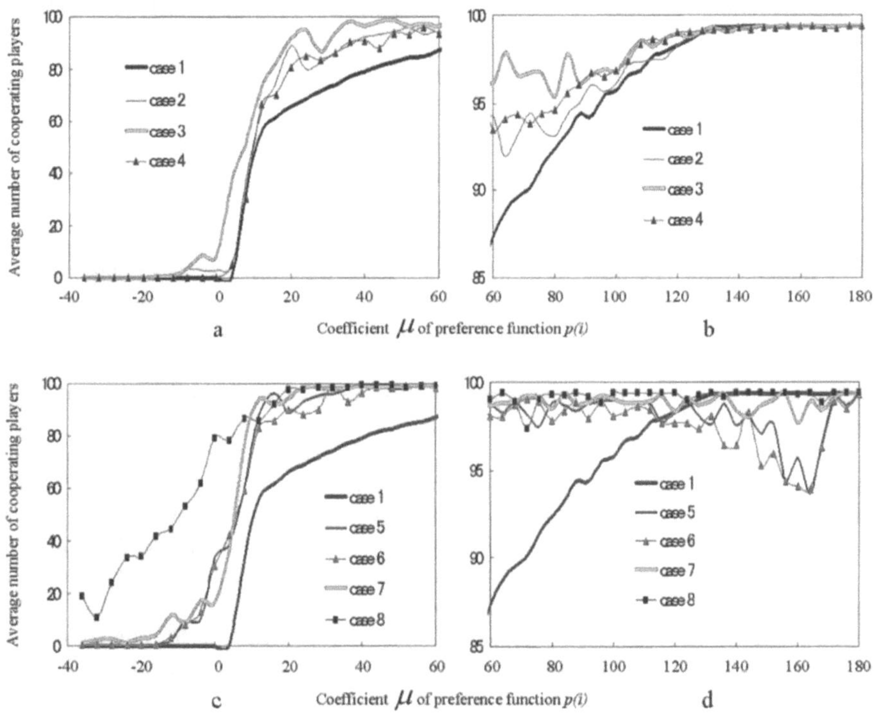

FIG. 1. Relations between the average number of cooperating players and coefficient μ in preference function $p(i)$. **a** Transitions in cases 1–4 when μ is –40 to 60. **b** Transitions in cases 1–4 when μ is 60 to 180. **c, d** Transitions in cases 1, and 5–8 when μ is 60 to 180

TABLE 2. Common parameters in the simulations in cases 1–8

Number of players	$\lvert N \rvert = 100$
Number of iterations	$T = 30\,000$
Maximum penalty	$Pen_{max} = 160$
Maximum tax	$Tax_{max} = 40$
Memory weight	$\delta = 0.9$
Coefficient of payoff function	$B = 160$
Coefficient of payoff function	$D = 2$
Coefficient of payoff function	$\alpha = 400$
Coefficient of normal distribution	$A = 4.54$
Coefficient of normal distribution	$\sigma = 10^{\frac{1}{2}}$
Strategy set of taxes $\Omega_{Tax_j} = \{0, 5, 10, 15, 20, 25, 30, 35, 40\}$	
Strategy set of penalties $\Omega_{Pen_j} = \{0, 16, 32, 48, 64, 80, 96, 112, 128, 144, 160\}$	

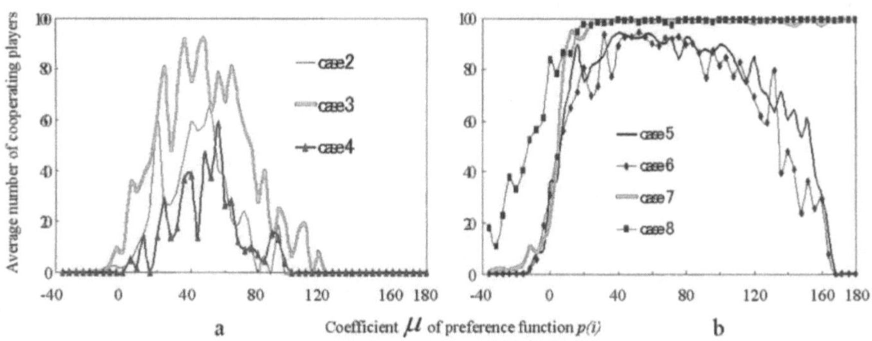

FIG. 2. Relations between the average number of cooperating players in a game with a metaplayer and coefficient μ in preference function $p(i)$. **a** Transitions in cases 2–4. **b** Transitions in cases 1, and 5–8

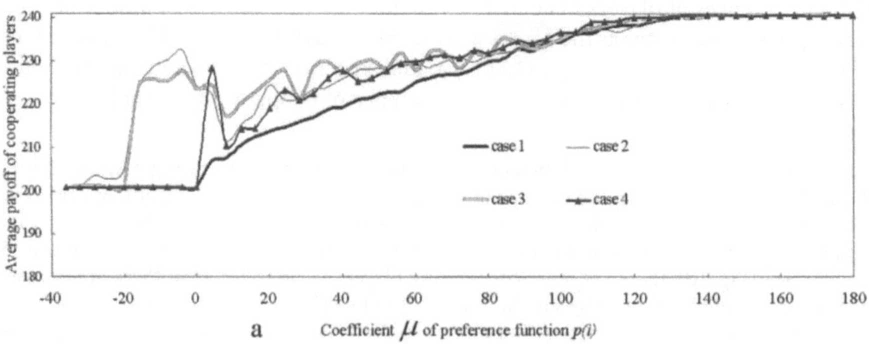

FIG. 3. Relations between average payoff for cooperating players and coefficient μ. **a** Transitions in cases 1–4. **b** Transitions in cases 1, and 5–8

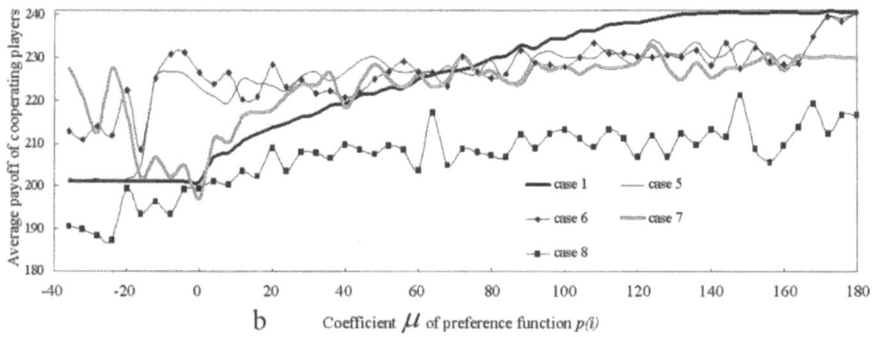

FIG. 3. *Continued*

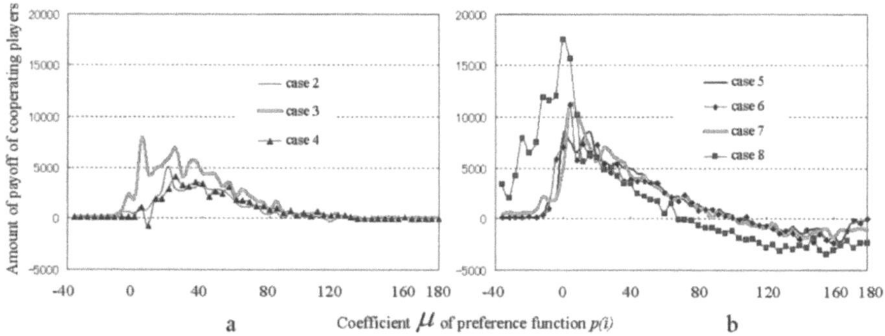

FIG. 4. Relations between amount of payoff for cooperating players and coefficient μ. **a** Transitions in cases 2–4. **b** Transitions in case 5–8

5.1 Number of Cooperating Players[1]

If the preferences of players to cooperate are not high ($\mu \leq 120$), the number of cooperating players increases as the games without metaplayers decrease. If the preferences of players to cooperate are high ($120 \leq \mu$), the existence of only one game without a metaplayer causes a negative effect in the number of cooperating players.

[1] First, we evaluated the effects of metaplayers on the number of cooperating players. Figure 1 shows the relations between the number of cooperating players and coefficient μ of preference function $p(i)$. Figure 2 shows the relations between the number of cooperating players in a game with a metaplayer and coefficient μ. Except for cases where coefficient μ is between 120 and 180 in cases 5–7 (Fig. 1d), there are more cooperating players in cases 2–8 than there are in case 1 (Fig. 1a–c). A comparison between the effects of competitive and cooperative metaplayers, showed that the number of cooperating players in case 3 is greater than in case 4 (Fig. 1a,b). The numbers of cooperating players in cases 5 and 6 are almost the same from the first game to the last (Fig. 1c,d). The number of cooperating players in case 7 is more than in case 8 until coefficient μ is 8. After that, the numbers of cooperating players in cases 7 and 8 are equal (Fig. 1c,d). If there are two metaplayers, competition between the two metaplayers increases the number of cooperating players, but competition among metaplayers assigned to all games decreases the number of cooperating players.

5.2 Average Payoff for Cooperating Players[2]

If the preferences of players to cooperate are low ($\mu \leq 80$), the average payoff for cooperating players increases with increases in competition among metaplayers or with increases in the number of games without metaplayers. However, if the preferences of players to cooperate are high ($80 \leq \mu$), the introduction of metaplayers does not increase the average payoff of cooperating players. An increase in metaplayers results in a decrease in the average payoff for cooperating players.

5.3 Amount of Payoff for Cooperating Players[3]

If the preferences of players to cooperate are very low ($\mu \leq 0$), the situation in which metaplayers are assigned to all games and they are cooperative is the most effective for increasing the amount of payoff for cooperating players. An increase in the number of games with metaplayers results in an increase in the amount of payoff for cooper-

---

The reason for the decrease in cooperating players in cases 5 and 6 when coefficient μ is between 120 and 180 is explained below. In this coefficient range, most players decide to cooperate even if there are no metaplayers because players lose the incentive to free ride (defect) if coefficient μ is greater than 160. However, there are still some players with a low preference to cooperate because the preferences of players to cooperate are based on a normal distribution. According to the conditions in cases 5 and 6, there is one game without a metaplayer. When coefficient μ is between 120 and 180, most players select the game without a metaplayer because there will be no tax (Fig. 2). The players with a relatively lower preference to cooperate decide to defect because they can always free ride. If there are some games without a metaplayer, cooperating players can avoid free riders by moving to another game without a metaplayer, such as those in cases 2–4.

[2] Next, we evaluated the effects of metaplayers on the average payoff for cooperating players. Figure 3 shows the relations between the average payoff for cooperating players and coefficient μ. According to the parameter settings, without the tax, the maximum payoff for one cooperating player is 240, and the minimum payoff is 200. Until coefficient μ reaches 80, the average payoffs for cooperating players in cases 2–7 are greater than in case 1. After that, the average payoffs for cooperating players in cases 2–4 are the same as in case 1. The average payoffs for cooperating players in cases 5–7 are less than in case 1, and the average payoff for cooperating players in case 8 is always less than in case 1.

As a result of a comparison between the effects of competitive and cooperative metaplayers, the average payoff for cooperating players in case 3 is the same as in case 4 (Fig. 3a). Similarly, the average payoff for cooperating players in case 5 is the same as in case 6 (Fig. 3b). The average payoff for cooperating players in case 7 is more than in case 8 (Fig. 3b). Thus, the effects of competitive metaplayers are only apparent in a comparison of cases 7 and 8.

[3] Finally, we evaluated the effects of metaplayers on the amount of payoff for cooperating players. Figure 4 shows the relations between the average payoff for cooperating players and coefficient μ. The increase in the amount of payoff for cooperating players is calculated by subtracting the product of the number of players and the average payoff for cooperating players in case 1 from that in cases 2–8. As a result of a comparison between the effects of competitive metaplayers and cooperative metaplayers, the amount of payoff for cooperating players in case 3 is more than in case 4 until coefficient μ reaches 100. After that, the amounts of payoff in cases 3 and 4 are equal (Fig. 4a). The amount of payoff for cooperating players in case 5 is the same as in case 6 from the first to the last game (Fig. 4b). Until coefficient μ reaches 20, the amount of payoff for cooperating players in case 7 is more than in case 8. After that, the amount of payoff for cooperating players in case 8 is more than in case 7 (Fig. 4b). The effects of competitive

ating players. If the preferences of players to cooperate are medium ($0 \leq \mu \leq 80$), the amount of payoff for cooperating players increases as the number of games with metaplayers increases. However, there are no large differences in these effects. If the preferences of players to cooperate are high ($80 \leq \mu$), the amount of payoff for cooperating players in all cases can be classified into two types based on the number of games with metaplayers. If there are only one or two games with metaplayers, the amount of payoff for cooperating players is zero regardless of whether there is competition or cooperation between metaplayers. However, if there are three or four games with metaplayers, the amount of payoff for cooperating players is negative. Furthermore, contrary to the case in which the preferences are very low, the situation in which metaplayers are assigned to all games and they are cooperative causes decreases in the amount of payoff for the most cooperating players.

According to comparisons between competitive and cooperative metaplayers, competition among metaplayers is not always effective in increasing the effects of metaplayers, particularly when there is a low preference of players to cooperate. A comparison of the effects of metaplayers showed that metaplayers who are suited to the preference of players to cooperate are needed to enhance the effects of an increase in numbers, the average payoff, and the amount of payoff for cooperating players. Metaplayers who are not suited to the current preference often have a negative effect.

In evolutionary contexts, it will be interesting to use the dynamic preferences of players to cooperate to replace static preferences. Furthermore, in order to construct a more realistic and general model, it may be useful to consider the cost of the introduction of metaplayers, and the penalty limited by the amount of tax (e.g., if there is no tax, metaplayer cannot impose a penalty).

6 Conclusion

We have proposed a game "Distributed social dilemma with meta-players" that is formulated using game theory. In order to determine the effects of metaplayers, we introduced the preferences of players to cooperate and the tendency of players to cooperate. The effects of metaplayers were evaluated from three aspects, i.e., the number, the average payoff, and the amount of payoff for cooperating players. A comparison of the effects of metaplayers showed that competition between metaplayers that is suited to the preference of players to cooperate is needed to enhance the effects of metaplayers.

---

metaplayers are apparent in comparisons between cases 3 and 4 and cases 7 and 8. A comparison between cases 3 and 4 shows that the presence of competitive metaplayers increases the amount of payoff for cooperating players. In a comparison between cases 7 and 8, the presence of competitive metaplayers reduces the amount of payoff for cooperating players until coefficient μ reaches 0. After that, the presence of competitive metaplayers increases the amount of payoff for cooperating players more than that in the presence of cooperative metaplayers.

References

Ashlock D, Smucker S, Stanley A, et al. (1996) Preferential partner selection in an evolutionary study of the prisoner's dilemma. BioSystems 37:269–300
Axelrod R (1984) The evolution of cooperation. Basic Books, New York
Axelrod R (1986) An evolutionary approach to norms. Am Polit Sci Rev 80:1095–1111
Axelrod R (1997) The complexity of cooperation. Princeton University Press, Princeton
Dawes RM (1981) Social dilemmas. Ann Rev Psychol 31:169–193
Epstein JM, Axtell R (1996) Growing artificial societies. MIT Press, Cambridge
Namatame A (1999) Multi-agents and complex theory. Morikita Shuppan, Tokyo
Schüssler R (1989) Exit threats and cooperation under anonymity. J Conflict Resolut 33:728–749
Tesfatsion T (1998) Gale-Shapley matching in an evolutionary trade network game. ISU Economic Report 43
Umino M (1989) Formulation of commons dilemma: Dawes model reconsidered. J Math Sociol 14:237–246
Uno K, Namatame A (1998) An evolutionary design of commitment networks. Proceedings of the 4th International Symposium on Artificial Life and Robotics, vol 2. pp 528–531
Yamagishi T, Hayashi N (1996) Selective play: social embeddedness of social dilemmas. In: Liebrand W, Messick D (eds) Frontiers in social dilemmas research. Springer, Berlin, Heidelberg, New York, pp 337–362
Yamashita T, Suzuki K, Ohuchi A (2000) The consideration of behavior process of players in social dilemma of iterated multiple lake game. J Soc Instrum Control Eng 36:195–203
Yao X, Darwen P (1994) An experimental study of *N*-person iterated prisoner's dilemma games. Infomatica 18:375–450

Part V New Frontiers for Evolutionary Economics

U-Mart: A Test Bed for Interdisciplinary Research into Agent-Based Artificial Markets

HIROSHI SATO*, YUHSUKE KOYAMA†, KOICHI KURUMATANI‡, YOSHINORI SHIOZAWA§, AND HIROSHI DEGUCHI†

Summary. U-Mart is a research project to construct a virtual stock market in which any preregistered agent can participate via networks. The aim of the project is to provide a forum where many agents of both computer programs and human traders participate in an auction and compete in a market, and also to provide a forum where research on market structure analysis is carried out. In this chapter, we introduce the concept and the system specification of U-Mart, and describe its importance and research possibilities in the context of economics, computer sciences, and artificial intelligence.

Key words. Artificial market, Efficient market hypothesis, Multi-agents, U-Mart, SVMP

1 Introduction

In this project, we provide a server service by preparing a "U-Mart server" that accepts inquiries and orders from "U-Mart clients" via the Internet. The communication between the server and the clients is done in simple virtual market protocol (SVMP) (details in Sect. 4).

The U-Mart server works in the same way as a real stock exchange market, i.e., collecting orders by agents, deciding the price, and paying/receiving the price to/from agents, except for the fact that real cash is not moved, i.e., the results of trading are just registered in the server, and that trade histories and final results will be open to all participants. In other words, the aim of the project is to provide a research forum for both computer science and economics researchers to compete in improving programming and economic analysis skills in the market.

In addition to this, we describe some possible research topics that are expected to emerge in the project, in the context of computer science and artificial intelligence as well as economics.

---

* National Defense Academy, Hashirimizu 1-10-20, Yokosuka-shi, Kanagawa 239-8686, Japan
† Kyoto University, Yoshida-Honmachi, Sakyo-ku, Kyoto 606-8501, Japan
‡ The National Institute of Advanced Industrial Science and Technology, Aomi 2-41-6, Koto-ku, Tokyo 135-0064, Japan
§ Osaka City University, Sugimoto 3-3-138, Sumiyoshi-ku, Osaka 558-0022, Japan

2 U-Mart as an Objective for Multi-Agent Research—Comparison between U-Mart and RoboCup Soccer

From the viewpoint of computer science and artificial intelligence, U-Mart is an example of a standard problem which we seek to solve by developing and testing new software techniques and computer programs. In particular, it is an attractive problem for a multi-agent system that consists of autonomous intelligent agents.

U-Mart is, in essence, influenced by a successful research project known as RoboCup Soccer (Kitano et al. 1995, 1997), mainly with the idea that it encourages the fast development of new ideas and techniques, and provides an open forum where experimental computer programs can be tested against each other.

However, U-Mart has some special characteristics that conventional techniques have not confronted. We describe these characteristics by contrasting U-Mart with RoboCup Soccer.

1. In a market, there is obviously no particular person or program for any individual program to compete with. Each program competes with the market itself. Program developers should investigate the market as a single entity consisting of many kinds of agents, rather than as a person or a team, which would increase the complexity of the computations.
2. Any game played in a market is nonzero-sum. All players have the same simultaneous possibility of gain or loss.
3. A market is a real open environment that is frequently influenced by information from other social systems, in contrast to RoboCup Soccer which is a closed environment.
4. The duration of a game is longer. Usually it takes a couple of weeks, or a few months, or possibly even years to decide on the winners. This means that players must have a long-term perspective in addition to short-term interests that will only lead to short-term gain.

The comparisons between U-Mart and RoboCup Soccer are summarized in Table 1.

3 U-Mart as an New Research Program for Economics—Comparison Between U-Mart and Ordinal Economic Paradigms

Recently, economics has adopted various arguments from other sciences such as computer science, artificial intelligence, decision theory, and behavioral science. Economics is now changing rapidly. In this section, we introduce some existing arguments about the economics of the stock market.

3.1 *Efficient-Market Hypothesis*

Economists have regarded a stock market as one of the most efficient markets because prices are adjusted rapidly, huge numbers of transactions are processed quickly by computers, and market participants incur negligible costs to receive information

TABLE 1. Comparison between U-Mart and RoboCup

	Investigation object	Number of agents	Game type	Openness	Game duration
U-Mart	Market itself	Many more O (100)–O (10 000)	Nonzero-sum	Open	Long O (1 month)
RoboCup soccer	Opposite team	O (10)	Zero-sum	Closed	Short O (10 min)

about market circumstances. The standard, simple efficient-market theory assumes identical homogenous investors who share rational expectations of an asset's future price. In the neoclassical framework, the market is a description of an exogenous mechanism for selecting prices that equalize aggregate supply and demand.

We now consider the simplest model of the efficient-market hypothesis. Consider a market with a single security that provides a stochastic dividend sequence d_t, with a risk-free outside asset that pays a constant r units per period. Assuming that all agents form the same expectations of the price and dividend in the next period under the common current information I_t, and that they arbitrate perfectly, the price p_t at time t is given as $p_t = (1 + r)^{-1}E[p_{t+1} + d_{t+1}|I_t]$, where $E[\ ]$ is the expected value. The market clears at the rational expectation equilibrium price p_t given by successive substitutions: $p_t^* = \Sigma(1 + r)^{-1}E[d_{t+n}|I_t]$.

The only source of price fluctuation is the change in the common information, and the dividend policy when the interest rate is constant. On the other hand, because all agents form identical expectations of the price, there is no incentive to speculate, and the efficient-market theory cannot explain the complex behavior of the market that emerges from this interaction of heterogeneous agents, e.g., bubbles and clashes.

3.2 Some Anomalies in the Efficient-Market Hypothesis

There is controversy about whether the efficient-market hypothesis can explain the reality of the markets. Neoclassical economists tend to support it, but psychologists and behaviorists are skeptical. Thaler (1992) collected anomalies that throw doubt on the efficient-market hypothesis.

3.2.1 The January Effect

The average monthly return to small firms in January is much higher than in other months. Half of this excess return to small firms comes in January, in the first five trading days. The January effect returns are higher for small firms that had lost value during the previous year, and the excess returns in the first 5 days.

3.2.2 The Weekend Effect

Prices tend to rise on the last day of the week. The average return on Mondays is very small (–0.18 points), in contrast with the average return on Fridays (or Saturdays) of 0.12 points. Prices on Mondays rise all day. The price of smaller firms shows greater changes in this weekend effect.

3.2.3 Holidays

Dow Jones Industrial Average (DJIA) shows a high proportion of advances the day before any holiday.

3.2.4 Turn-of-the-Month Effects

All returns for a month occur in the first half of the month, and the returns for the second half of the month are negative.

3.2.5 Intraday Effects

On all days except Mondays, prices rise during the first 45 min. Returns are high near the end of the day, particularly for the last transaction of the day. The day-end price changes are greatest during last 5 min, and this has also been observed in experimental markets.

Thaler (1992) also said that structural and institutional reasons can explain some of these phenomena, for example, (1) the different lengths of closing times for different markets, and (2) the specific arrangement made for a settling day. However, we need other approaches to model the phenomena in order to explain them totally.

3.3 Noise Trader Approach

There are various types of trader in the real market. From their motivations to deal, traders are classified under three types, the informed traders, the uninformed traders, and the market makers.

The informed traders have some information that is based on economic fundamentals and is not common to market participants, and they use this when dealing to maximize their profits.

Some uninformed traders have neither information nor an interest in speculation. They trade for liquidity, and purchase at the asking price or sell at the lid price, depending on their estimates of the asset value of liquidity requirements. These are called liquidity traders. The other uninformed traders also have some information, which is not based on economic fundamentals but on technical analyses, i.e., market trends. Technical analysis typically calls for buying more stock when stocks have risen (breaking through a barrier), and selling stocks when they fall through the floor. Both liquidity traders and technical analysts are often called "*noise traders*" (Shleifer and Summers 1990).

Market makers must deal with informed and uninformed traders. They are used as market intermediaries whose main function is to figure out ways of clearing the market, e.g., pricing to match purchases to sales.

Some market participants have expectations based on market trends in the real market, which means that the investors can follow the technical analysts and make use of the fluctuations of the market in order to make more profit. An investment plan based on the economic fundamentals is not always profit-maximizing behavior. It is important for investment success not only to predict future fundamentals precisely, but also to predict the movement of other active investors.

Trend-chasers' behavior, i.e., buying after stock prices rise and selling after they fall, amplifies the fluctuations of the market, since they follow positive feedback strategies.

3.4 Individual Decision Making Under Risk and Uncertainty

Since Allais (1953), it has been known for more than forty years that individual decision makers do not behave in accordance with the axioms of expected utility theory. In addition, because what kind of fundamental information should be noted, or how information should be decoded, have not been completely settled, traders' decisions about selling and buying depend on their cognitive frameworks and sentiments, and these are the research areas of cognitive psychologists and decision theorists.

3.4.1 Scenario Thinking

Tversky and Kahneman (1973) highlighted a common cognitive error, known as *scenario thinking*, that guesses at the probability of an occurrence are influenced by the ease with which one can imagine a phenomenon, and this they called the "bias of imaginability." Although it is necessary in cases where there is little information about the future, it must be recognized that scenario thinking sometimes extrapolates excess optimism or pessimism about the market.

3.4.2 Prospect Theory

Kahneman and Tversky (1979) extended the expected utility theory and described a model of decision making under risk and uncertainty. The individual's evaluation function is (1) defined by deviations from the reference point, (2) generally concave for gains and commonly convex for losses, and (3) steeper for losses than for gains, e.g., risk-averse. The decision weight function $\pi(p)$ of probability p is (1) generally lower than the corresponding probabilities, (2) except in the range of low probabilities. Overweighting the low probabilities may contribute to the attractiveness of speculative trading. In addition, as prices are always evaluated from relative changes, the sequential movement of prices may influence the individual's judgment of purchases and sales.

3.5 Artificial Stock Market

Arthur et al. (1997) proposed the artificial stock market (ASM) approach, where all agents are bound to be rational and form price expectations like uninformed human traders. Each agent has 100 predictors that are used for genetic algorithm (GA) learning for technical trading, and 25 agents which participate in the artificial market.

Izumi and Ueda (1998) integrate the fieldwork and the multi-agent model. They interviewed a dealer and found that the features of dealer interactions in learning were similar to the features of genetic operations in biology. They constructed an artificial market model with 100 agents using a GA; each agent has a prediction method which imitates that of a dealer. Their model also showed several emergent phenomena: the peaked and long-tailed distributions of rate changes, the negative correlation between trading amounts and fluctuations, and the contrary opinion phenomena resulting from phase transition.

TABLE 2. Comparison among research approaches to market microstructure

	Agent type	Parameters	Openness	Complex behaviors
Efficient market hypothesis	Homogenous	Fixed	Closed	Impossible
Noise trader approach	Heterogeneous	Fixed	Closed	Possible
Artificial stock market	Homogenous	Learned	Closed	Possible
U-Mart	Heterogeneous	Any type	Open	Possible

3.6 U-Mart as a Hybrid Approach

These ASMs can express various kinds of market phase. However, agents have a preprogrammed capability in technical trading, and the period of the moving average is fixed, which might cause systemic oscillations. After all, there are limits to the types of agent that the ASM approach can express.

One of the important characteristics of U-Mart is its openness. This consists of public participants preregistered to the market. The market is expected to consist of two types of agent, i.e., machine automatic agents and human agents. Because there always exist various types of price expectation that bring diversity to the market, we can investigate complex interactions between machine and human agents (Table 2).

4 The U-Mart Experimental System

In this section, we introduce the artificial market simulator system and the communication protocol that were developed by the U-Mart organizing committee.

4.1 System Specification

Figure 1 shows the composition of the U-Mart experimental system. The system consists of a virtual market server and trader clients. The clients communicate with the server by simple virtual market protocol (SVMP). SVMP is a character-based protocol on TCP/IP. This means that the server and the clients can be developed separately, and the program is platform-free. The U-Mart server contains a communication component and a database component, in addition to a transaction component. SVMP is used for communication and PostgreSQL is used for the database. The system is implemented by Java language because Java is easiest for both network and database programming.

4.1.1 Transaction Specification

U-mart adopts interval time trading; orders are accepted at any time, but contracts are enforced at each discrete time step. This is because of the delays on the Internet.

The traders have to be members of the market. Some virtual money is given to each member for trading. The order of trading is controlled by consideration of the cash and position of each member in order to prohibit reckless trading.

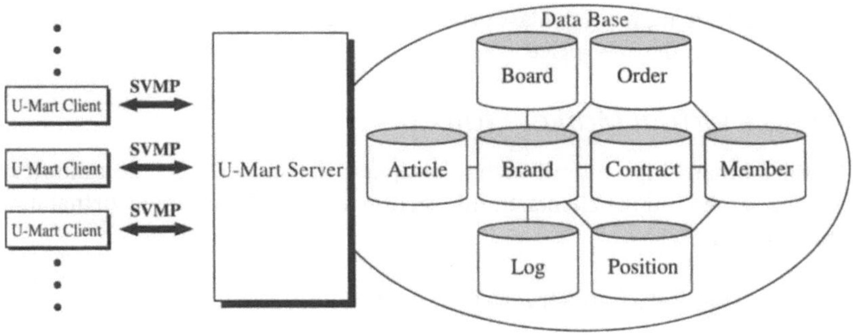

FIG. 1. Composition of the U-Mart experimental system

When each virtual trading day is finished, the market calculates the property of the members by speculation from a spot price. If the capital of a member is less than a preset criterion, they go bankrupt and cannot continue trading.

4.1.2 Articles Specification

The U-Mart server deals with virtual futures of the real stock index. This establishes a one-way connection to the real world. We are listing futures of J30, the stock exchange index of 30 representative Japanese companies. The detailed information is as follows:

— article, J30;
— delivery time, 3 months;
— final settlement day, next business day after final trade day;
— method of final settlement, speculating for the difference from the spot price;
— maximum price change limit, none.

4.1.3 Transaction Specification

There are two kinds of order: order and cancel order. An order has five attributes and cancel order has three attributes, as shown below.

1. Order:
   — order ID, added by server (integer);
   — new or reverse (1, new; 2, reverse);
   — sell or buy (1, sell; 2, buy);
   — market price or limit price (1, market; 2, limit);
   — volume (integer).
2. Cancel order:
   — order ID (1), added by server (integer);
   — order ID (2), that you want to cancel (integer);
   — volume (integer).

The server holds all the contracts for the new position and balances them for an existence position. Since we deal in stock index futures, settlement is done by speculating on the difference from the last price. Final settlement is done by speculating on the difference from the spot price.

4.2 Simple Virtual Market Protocol

The U-Mart organizing committee has designed the protocol for communication between the server and the clients. In this section, we introduce simple virtual market protocol (SVMP).

4.2.1 Connection Between Server and Clients

The server and the clients communicate through TCP/IP using a socket interface. The port number for U-Mart is 5010. The status of the connection is either connected or nonconnected. First, the trader sends their ID and password to the server, and after certification, the transaction begins.

4.2.2 Format of Message

Messages are divided into the following three types: (1) prompt (server → client), (2) request (client → server), and (3) result (server → client). Below is a brief specification of the message format.

[available character] ASCII characters
[word] one or more characters divided by delimiter
[delimiter] space or tab
[maximum length] line, 15 words; word, 64 characters
[date and time] YYYY/MM/DD, HH:MM:SS
[prompt] before certification, "+LOGIN"; after certification, "+OK"
[request] instruction argument1 argument2 . . .
[result] normal condition, "+ACCEPT"; illegal condition, "+ERROR"

4.2.3 Commands

Table 3 shows the main commands of U-Mart. Commands are classified into the following four categories: (1) certification, (2) status of server, (3) orders, and (4) inquiries. There are other commands for the administrator only.

4.3 Tools

Basically, the server is a character-based program, but graphical components can be attached. If we execute U-Mart with graphical components, the parameters are set through a window (Fig. 2), and the results are shown in charts in another window (Fig. 3).

The U-Mart experiment system includes three types of sample agent and a graphical user interface (GUI) for human traders. Sample agents are: (1) a random trader who attempts to buy and sell randomly, (2) a trend trader who attempts to buy if the price leaps, and sell if the price drops, and (3) an antitrend trader who is the exact opposite of the trend-trading agent. The volume of their transactions is determined at random.

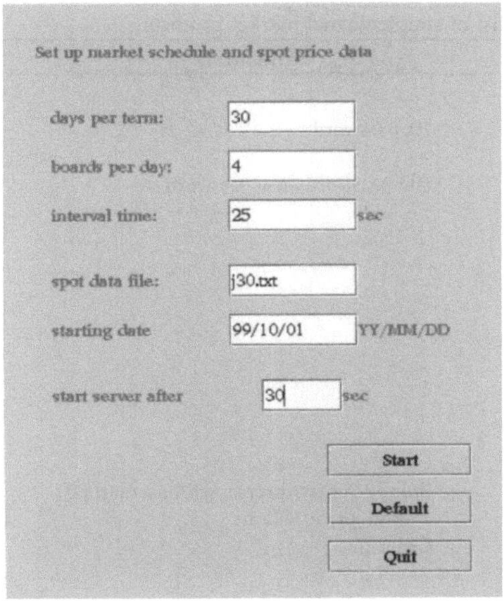

FIG. 2. Parameter window for the U-Mart server

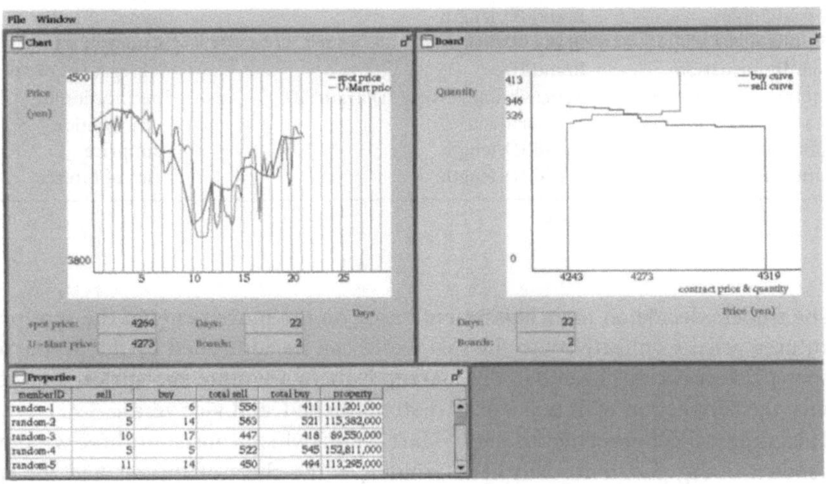

FIG. 3. Graphical user interface for the U-Mart server. *Top left-hand* window is a price chart: spot price (*solid line*) and futures price (*dashed line*). *Top right-hand* window is a contract chart: buy curve (*solid line*) and sell curve (*dashed line*). *Bottom* window shows the position list for all traders

TABLE 3. Command list of simple virtual market protocol

Command	Arguments	Function
Certification		
Login	ID, password	Logging in
Logout	–	Logging out
Passwd	Old password, new password	Change password
Status, of server		
Schedule	–	Trade schedule
StartTime	–	Start time
MarketStatus	–	Status of server
ProtocolVersion	–	Version of SVMP
IntervalTime	–	Interval of contract
ServerTime	–	Time of server
ServerDate	–	Date of server
Order		
OrderRequest	Brand ID, new/reverse, sell/buy, market/ limit, price, volume	Order
OrderCancel	Order ID	Cancel order
Inquiry		
OrderStatus	–	Inquiry about order
Executions	–	Inquiry about contract
Position	–	Inquiry about position
Balances	–	Inquiry about balance
TodaysQuotes	Brand ID, length	Quotes of today
HistoricalQuotes	Brand ID, length	History of quotes
BoardInformation	Brand ID	Aggregate of orders
ContractSpecification	Article/brand, article ID/brand ID, assortment	Specification of transaction
SpotPrice	Brand ID, length	Spot price
FuturesPrice	Brand ID, length	Price of futures

The traders decide on their investment based on the movements of the spot price or futures price. Connectivity to the real world can be controlled by the proportion of spot-price-conscious agents in the market. If there are more spot-price-conscious agents, the connection with real stocks is strengthened, and vice versa.

The graphical components for the U-Mart server, which consist of three windows, are shown in Fig. 3. The top left-hand window is the chart of futures and the spot price. The top right-hand window depicts the supply and demand lines. The point of intersection shows the contract price. The lower window shows the properties of all the traders.

Figure 4 shows the GUI for human traders. The GUI has four windows. The top left-hand window shows the property of the trader. The top right-hand window shows the price chart. The bottom left-hand window follows the time sequence of the trader's profit. The trader can request or cancel an order using the bottom right-hand window.

At present, the U-Mart server can establish hundreds of connections and accept thousands of orders each second.

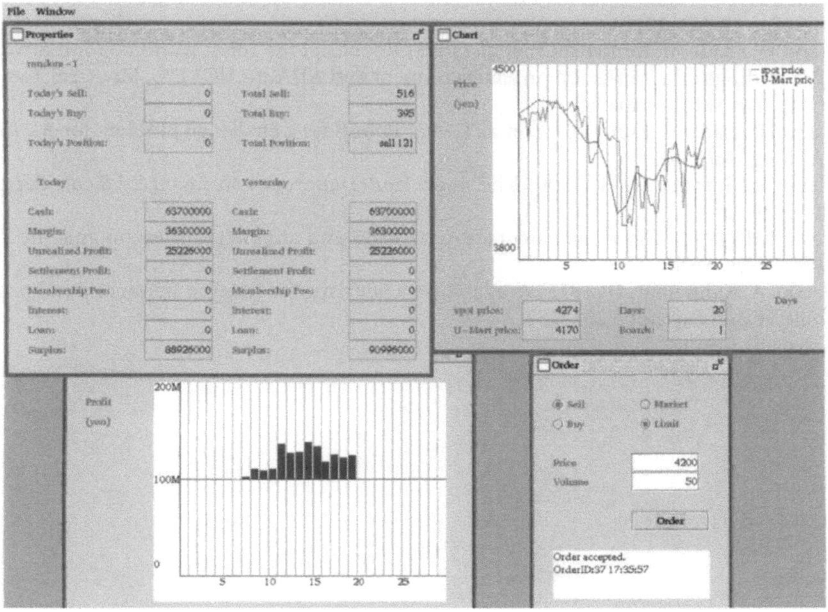

FIG. 4. Graphical user interface for a U-Mart client. The *top left-hand* window shows information about the client's position. The *top right-hand* window is a price chart: spot price (*solid line*) and futures price (*dashed line*). The *bottom left-hand* window shows the time evolution of the client's profit. The *bottom right-hand* window is an order form

5 Conclusion

We have introduced the outline of a research project known as "U-Mart," which is a forum for multi-agent system research, and a research tool for market structure analysis. The U-Mart committee distributes the U-Mart training kit from the U-Mart web page (http://u-mart.econ.kyoto-u.ac.jp/). Any researchers or actual traders are welcome to the market in order to prove and improve their programming techniques and trading skills.

References

Allais M (1953) Le compartement de l'Homme Rational devant la Risque, Critique des Postulats et Axiomes de l'Ecole Americane. Econometrica 21:503–546

Arthur WB, Holland JH, LeBaron B, et al. (1997) Asset pricing under endogenous expectations in an artificial stock market. SFI studies in the sciences of complexity, vol XXVII. Addison-Wesley, Reading

Izumi K, Ueda K (1998) Emergent phenomena in a foreign exchange market: analysis based on an artificial market approach. In: Adami C, Belew R, Kitano H, Taylor C (eds) Artificial life, vol VI. MIT Press, Cambridge, pp 398–402

Kahneman D, Tversky A (1979) Prospect theory: an analysis of decision under risk. Econometrica 47:263–291

Kitano H, Asada M, Kuniyoshi Y, et al. (1995) RoboCup: the robot world cup initiative. In: Proceedings of a Workshop on Entertainment and AI/Alife, IJCAI'95, Moutreal, Canada (http://www.robocup.org)

Kitano H, Asada M, Kuniyoshi Y, et al. (1997) RoboCup: a challenge problem for AI. AAAI AI Mag 18:73–85

Shleifer A, Summers, LH (1990) The noise trader approach to finance. J Econ Perspect 4(2):19–33

Thaler RH (1992) The winner's curse—paradoxes and anomalies of eeconomic life. Free Press, New York

Tversky A, Kahneman D (1973) Availability: a heuristic for judging frequency and probability. Cogn Psychol 5:207–232

Econophysics: Empirical Laws, Theory, and Application

Hideki Takayasu

Summary. Analyses of financial data using the methods of statistical physics have revealed the existence of many nontrivial statistical laws. A typical example is the fractal properties found in the price fluctuations of any open market, which can be understood theoretically by introducing the concept of phase transitions for the market between excess demand and excess supply. The critical-point behavior can be described by a Langevin-type equation with a fluctuating coefficient, and the basic properties of price fluctuations can be explained. As an application of the econophysics approach to the foreign exchange market, a new electrical money system is introduced that can protect an international company from turbulent market fluctuations.

Key words. Econophysics, Power law, Fractal, Phase transition, Electrical money

1 Background

The new word "econophysics" first appeared in 1997 at an international workshop on econophysics held in Budapest. Many physicists gathered at this workshop, and they confirmed the existence of nontrivial empirical laws that can be established by methods developed in the field of statistical physics. Since then, this emerging field of science has been attracting increasing numbers of researchers not only from the field of physics, but also from economics and finance. Econophysics is a link between economics and physics in the way that biophysics connects the fields of biology and physics (Mantegna and Stanley 2000).

The characteristics of econophysics can be described by the keyword "fluctuations." In ordinary economics, which is based on the concept of a balance between supply and demand, fluctuations tend to be neglected. A typical example is the spring model of supply and demand which can be found in any textbook of economics. Such a system has no fluctuations, and can only be realized in physical systems at the zero temperature limit. On the other hand, in the framework of financial technology,

Sony Computer Science Laboratories, Higashigotanda 3-14-13, Shinagawa-ku, Tokyo 141-0022, Japan

fluctuations are dealt with as noncorrelated Gaussian noise or purely mathematical stochastic processes. Such a stochastic process is an idealized model for real physical systems that are valid to the limit of infinite temperature. In econophysics, fluctuations are considered to have both stochastic and deterministic properties, like all complicated systems in the real world of finite temperature.

Intensive studies on finite-temperature phenomena in statistical physics have developed many tools and concepts in the last 40 years. Chaos, phase transition, and fractals are typical keywords in this field of science. The merit of econophysics comes from using these new tools and concepts in physics to analyze financial data and numerical models, as described in the following sections.

In the second section, some data analysis results are summarized both for market price fluctuations and for companies' financial data. In the third section, the phase transition view of the market is introduced, and a set of equations des cribing the price dynamics is derived. As an application of econophysics, we discuss risk hedge methods for understanding market price fluctuations in foreign currency exchanges, and as an ultimate solution we introduce an electrical money system in the fourth section. The final section is devoted to discussions of the goal of econophysics.

2 Data Analysis

2.1 Open Market Price

From the viewpoint of statistics, econophysics is typically characterized by power-law distributions, while financial technology is based on normal distributions. In this sense, the root of econophysics can be found in the pioneering work done by Mandelbrot in the 1960s (Mandelbrot 1997). He analyzed the historical market price data of cotton and found that the price changes were scale-invariant both in the time scale and in the price scale. His discovery was checked against the high-precision database of the stock market by Mantegna and Stanley, who claimed that the average market price fluctuations in a fixed time interval follow a symmetric power law, with cutoffs for large values (Mantegna and Stanley 1995). It has recently been clarified that for each stock price fluctuation, the cumulative distribution of price differences follows a power law with an exponent of about −3 for wide time-interval scales (Gopikrishnan et al. 1998).

The foreign exchange market has also attracted much interest because of its peculiar characteristics. The market is open continuously for 24 h a day except at weekends, the total money flow around the globe is the largest of all the open markets and is estimated to be up to US$ 2×10^{12} per day, and there is no regulation such the circuit-breaker systems of the stock markets. Although there are interventions by central banks occasionally, the amount involved is rather small compared with the net flow, so the market is considered to be out of any person's control. In this sense, the rate fluctuation of the foreign exchange market can be regarded as a kind of spontaneous fluctuation.

Figure 1 shows an example of the yen–dollar rate fluctuation, with the time axis being plotted in ticks. Here, a tick is equivalent to the occurrence of a transaction, and

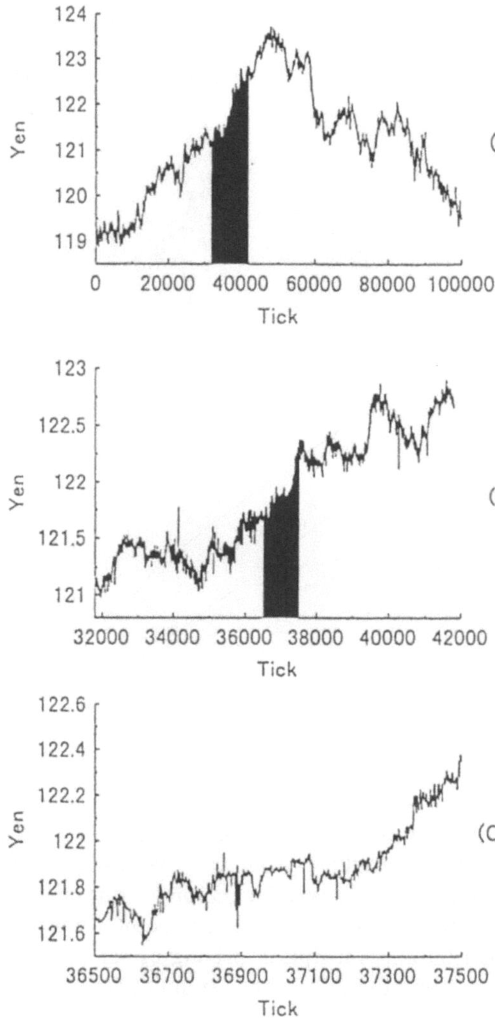

FIG. 1. An example of yen–dollar fluctuations. In each case, the *dark sections* are magnified 10 times in the figure immediately below. See text for an explanation of a "tick"

about 12 000 ticks are reported every weekday, i.e., a tick is about 7 s on average. A close-up of part of the fluctuation looks similar to the whole, and a further close-up also looks statistically quite similar, as shown in the figure. This self-similarity is the fractal property that Mandelbrot pointed out (Takayasu et al. 2000).

The probability density of exchange-rate change in a tick is shown in Fig. 2 on a semi-log scale. The dotted line shows the best-fit normal distribution which has the same mean and variance. It is clear that the observed distribution is far from the normal distribution, and is much closer to a power law. Of course this power law is bounded in the scale size; the small scale limit in the sense of statistics is the minimal transaction unit ¥0.01 and larger limit is estimated from the data to be about ¥1.0 for US$1.

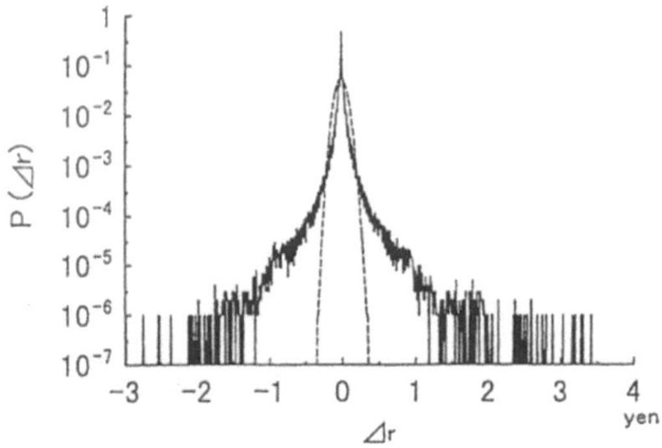

FIG. 2. Probability density of the yen–dollar rate difference per tick in a semi-log plot

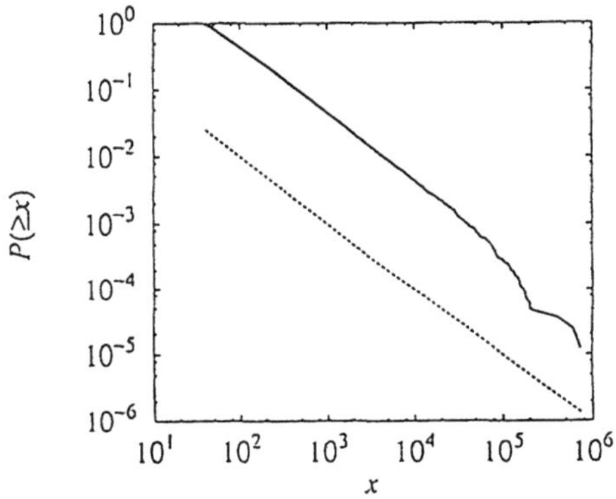

FIG. 3. Annual income distribution of Japanese companies in 1997

As seen from the examples shown here, the distribution of market price fluctuations is generally far from a normal distribution in any open market, and is generally closer to a power law. The reason for this power-law behavior is discussed in the third section from a theoretical view point. Applications of effects such as fat tails in the distributions for option pricing or for portfolios have been considered in various forms, and these have been summarized by Bouchaud and Potters (2000).

2.2 Company Data

Other important quantities in the study of econophysics are company data such as incomes and assets. These quantities also fluctuate, but the time scale of the fluctuation is much longer than in the case of open markets: instead of a few seconds for open markets, a typical time interval for company data is 1 year, i.e., more than a million times longer.

In Fig. 3, the distribution of the incomes of companies is plotted on a log–log scale for all Japanese companies that reported more than ¥4×10^7 income to the tax offices in the fiscal year 1997. Here, income means roughly total sales minus total expenses. It is clear that the plots lie on a straight line, with the exponent being very close to −1 except for those companies with very large incomes. Such a power-law distribution with exponent −1 is called Zipf's law. The same statistical law has also been confirmed for Italy, France, and several other countries (Okuyama et al. 1999). From these examples, it can be expected that Zipf's law for company incomes holds universally.

More than 100 years ago, Pareto investigated personal income distributions for several European countries and reported that the distributions were generally approximated by power laws (Pareto 1897). In the field of econophysics, personal income distribution is also attracting much interest.

Recently, high-accuracy personal income data from Japan have been analyzed by Aoyama et al. (2000), and they confirmed that the income distribution of the majority of tax payers is approximated by a log-normal distribution, while the richest people's incomes satisfy a power law. The exponent of this power law is very close to 2.0, which is different from the Zipf case where it is 1. No theory has yet been established which can explain the power laws of incomes.

3 Phase Transition View

During the intense period of research into fractals in the 1980s, the scale-invariant phenomena, characterized by power laws, were investigated in many interdisciplinary scientific fields and many theoretical models were proposed (Takayasu 1990). Of course, each model has its own individual characteristics, but it became evident that most of the models involved some kind of phase transition mechanism, either explicitly or implicitly.

In statistical physics, it is well known that fluctuation is greatest at the phase transition point, which is also called the critical point. The critical point is characterized by fractal properties such as self-similarity or power laws. In classical physical models, such as the models of magnets, the phase transition point is generally unstable, and the control parameter, e.g., temperature, must be regulated with great care in order to sustain the system at the critical point. On the other hand, some of the models of fractals showed that mechanisms exist that stabilize the critical point; this is called self-organized criticality, or SOC (Bak 1996). The models with SOC automatically show critical point properties without any controlling parameters, and in some models it has been rigorously shown that the system always converges to the critical point from any initial condition (Takayasu 1989).

An open market can be recognized as having a kind of SOC in the following sense. The balance of supply and demand is often understood by using the spring model in economics, although in the real world demand is more like the number of buyers in an open market, which fluctuates randomly, than like a spring force. This stochastic property is better understood by a phase transition approach: the excess-demand phase, in which the average demand is larger than the supply, and the excess-supply phase, where the opposite situation applies. According to this view, the balanced point is the phase transition point by definition (Takayasu and Takayasu 1999). In a pioneering work on artificial markets by Hirabayashi et al. (1993), it is shown that the market price fluctuates like a random walk, and even the underlying dynamics is deterministic if the numbers of buyers and sellers are balanced, while the resulting price shows a drift if the numbers are not balanced.

Can the balancing point really be viewed as a phase transition point? The known facts about fractal properties introduced in the previous section can be regarded as evidence, but they are not sufficient. In order to promote the phase transition view point, we now introduce a set of dynamic equations describing the price fluctuations around the critical point.

Let $p(t)$ and $p^*(t)$ be the market price and the balanced price at time t. The balanced price, $p^*(t)$, is not directly observable, since it can only be determined if all the traders show their intentions openly. As the market price tends to be attracted by $p^*(t)$, we have

$$p(t+\Delta t) - p(t) = -A(t)(p(t) - p^*(t)) + f(t) \tag{1}$$

where $f(t)$ is random noise, and the coefficient $A(t)$ is a quantity representing liquidity, or the inverse of price rigidity.

The value of $p^*(t+\Delta t)$ is determined in the traders' minds, which are expected to be responsive to the latest market price changes, and we then have

$$p^*(t+\Delta t) - p^*(t) = B(t)(p(t) - p(t-\Delta t)) + f^*(t) \tag{2}$$

where $f^*(t)$ is another random number, and $B(t)$ is a coefficient that represents the traders' tendency to follow the local trend in market price changes. This set of linear stochastic equations represents the market price dynamics (Takayasu and Takayasu 1999).

If both A and B are small, the market price fluctuation is very small. When B is large but A is small, an exponential price change occurs which is equivalent to a so-called "bubble." If A is large but B is small, then the market price shows a damped oscillation. When both A and B are large, the market prices diverge with large oscillations.

The power-law tails in the market price change already shown in Fig. 2 are understood theoretically from this set of equations in the following way. First, we apply the approximation $p(t) \sim p^*(t)$, which is expected to be reasonable if we observe the price changes on a large scale in both price and time. Then the set of price equations becomes the simple autoregressive-type stochastic equation

$$X(t+\Delta t) = B(t)X(t) + F(t) \tag{3}$$

where $X(t) = p(t) - p(t-\Delta t)$ represents the price change, and $F(t)$ is the random force.

From the viewpoint of physics, this equation is a discrete version of the Langevin equation which describes fluctuations due to thermal noise in general, such as in an electrical circuit. In thermal equilibrium, the coefficient B is a constant which is related to the random force $f^*(t)$, as is well known in the fluctuation dissipation theorem, and B cannot take a value larger than 1 because $B > 1$ implies a kind of amplification or instability. Here, however, we consider the case where $B(t)$ is also a random variable independent of $f^*(t)$, and also the case where $B(t) > 1$. The situation where $B(t)$ is fluctuating around $B(t) = 1$ is also consistent with the phase transition view because $B = 1$ corresponds to the phase transition point of stable and unstable phases.

In mathematics and in theoretical physics, the stochastic Eq. 3 is well known as multiplicative noise or random coefficient Langevin system (Takayasu et al. 1997). Passing over the details of the derivation, the main result of the theoretical analysis is summarized simply in the following way: starting from any initial condition, the system converges to a statistically steady state in which the distribution of $X(t)$ follows a power law

$$P(>|X|) \propto |X|^{-\beta} \tag{4}$$

where $P(>X)$ denotes the cumulative probability that a chosen sample is larger than X. The exponent β is shown to satisfy the following equation rigorously:

$$<B(t)^{\beta}> = 1 \tag{5}$$

By using this theorem, we now understand that the reason for the power law behavior of the price difference is the fluctuation of $B(t)$, the traders' average tendency to chase the latest price trend. In ordinary situations, the coefficient is less than 1 and the price fluctuation is dissipated, but occasionally traders become extra sensitive to price changes and $B(t)$ becomes greater than 1, resulting in large price differences. By assuming that such a fluctuation of $B(t)$ occurs randomly, we can say that the price difference distribution should follow the power law.

As an application of this theory, it has been demonstrated that in an electrical circuit with a mechanism of random amplification, the intrinsic noise deviates from Gaussian statistics but follows a power law (Sato et al. 2000). This result also suggests the close similarity of market price changes and noises in an electrical circuit.

The cutoff in the distribution for very large values is easily implemented in the theoretical approach by considering a kind of nonlinear effect such that $B(t)$ does not take a value larger than 1 when $|X(t)|$ is larger than a cutoff value X_c. This effect is also reasonable for traders, because many traders will feel that very large price changes do not continue, so there is a kind of nonlinearity in the response of $B(t)$.

As can be seen from these discussions, the open-market system has a mechanism that keeps the market price fluctuations close to the critical point. There have been optimistic expectations, based on the spring model of supply and demand, that the market price can be stabilized by reducing the restrictions and increasing the information. However, econophysics predicts the opposite scenario, i.e., by reducing restrictions and increasing information the system will move closer to the market's balancing point. This balancing point has critical-point characteristics, and therefore the price fluctuations will become even larger and quicker. This forecast seems to be compatible with the fact that the foreign exchange market did not become stable even after the global liberation of 1998.

4 A New Electric-Money System

As we have seen in the preceding section, the present market system of currency exchange has a systematic problem of accompanying critical fluctuations. There are ways to cope with such large fluctuations. One is to deal with the fluctuations by introducing insurances such as options and derivatives. Other approaches are to control the fluctuations by interventions, or to escape from exchange risk by introducing a new currency, such as the Euro. Here, we introduce a new currency system for international private companies which is aimed at avoiding the risks due to fluctuations of foreign exchange systems.

An international company is open to several different types of risk from fluctuations in foreign exchange rates. The most typical problem is a contradiction in the company accounts; for example, a company appears to be growing if we calculate the whole company's assets in dollars, but it is seen to be losing money if the calculations are done in yen. Another type of problem is pricing. For example, let us assume a situation where we have priced a commodity equivalently in both yen and dollars at a certain time following the exchange rate at that time. After a while, if the yen–dollar rate changes by 20%, then the values of the two prices in yen and dollars are no longer equivalent, but show a 20% difference. From the stand point of customers shopping via the Internet, it is very easy to choose the cheaper product. Such multiple pricing is practically equivalent to a discount, and therefore the exchange rate fluctuations can be a big problem for a company.

Can we reduce such risks? In the sense of econophysics, an analogy between the market and an electrical circuit can be applied. Reducing the market fluctuations is similar to finding a solution for reducing the noise in an electrical circuit, because both systems can be described by similar dynamics.

If we are making an electrical circuit, what should we do to reduce the noise level? This is a well-established problem in physics, but there is an answer. First, exclude the parts from which a lot of noise is generated. Second, introduce a noise cancellation mechanism by amalgamating the noise in both space and time. We can apply this idea by introducing a new currency system which is applicable to an international private company suffering demerits or risks from fluctuations in major currencies.

Speculators in a currency market can be regarded as noise sources in an electrical circuit because large fluctuations with short time scales are created or enhanced. It is impossible to exclude speculators from an existing currency market, but if a company introduces its own currency system, with private company money, then speculative trade can be excluded or prohibited with respect to this new currency.

The new currency is an electrical money transferred digitally by a computer network. Orders of exchange to existing currencies are gathered in a central banking system that processes all tasks about this currency. The central bank keeps a certain amount of money in existing currencies which works as a buffer for exchanges of company money and ordinary currency. Therefore, the company money is exchangeable into any currency instantaneously via the computer network. As all orders are centrally processed using the buffer, fluctuations are averaged over space and time.

The key point of the new currency system is how to determine the exchange rate. Ideally, the exchange rate should be determined so that the company's account is independent of the fluctuations in the exchange rates of existing currencies. This ideal situation can be realized by the method described below.

Let us consider a simple case where a company has assets in two currencies, say, x dollars and y yen. When the exchange rate is \$1 = ¥r, the total assets estimated in yen are $A_y = y + rx$, and the assets estimated in dollars are $A_d = x + r^{-1}y$, neglecting the rate difference in the opposite direction. When the value of r changes, the estimated asset values change accordingly even if the assets themselves do not change. This problem can be avoided if we define the new currency unit as being proportional to (x dollars + y yen), as shown schematically in Fig. 4. In this figure, the dashed line shows the constant value combination of dollars and yen in the case of an exchange rate r, and it is clear that the point (x dollars + y yen) is fixed for any value of r. In other words, the new currency should be a kind of basket of dollars and yen in a ratio which is proportional to the company's total assets. The company's account then becomes

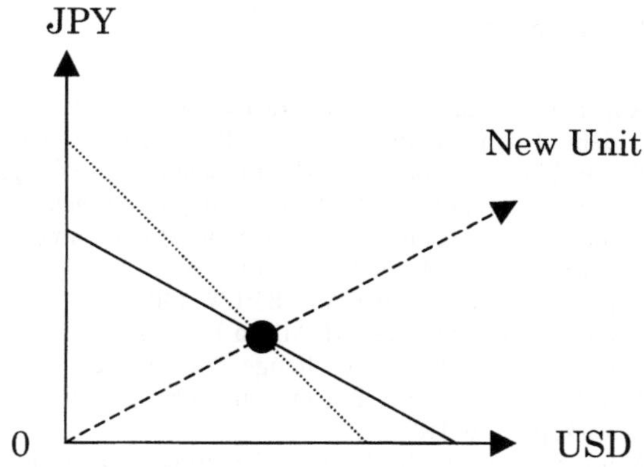

FIG. 4. Asset configuration (the *large dot*) in a yen–dollar plot. Even when the exchange rate changes from the *solid line* to the *dotted line*, the new unit of money is invariant

invariant when there are changes in r. This method can also be applied to the case of many existing currencies.

5 Discussions

Although money appears to be additive and linear, the functions or values of money are definitely nonlinear. For example, consider a figure of one million dollars. If this amount of money is saved in a bank account, it grows by interest. However, if the money is distributed to one million people who get one dollar each, the money does not grow at all. In other words, the large amount of money has the property of self-growth and its value is larger. This is the basic reason that money tends to accumulate.

Traditionally, it is said that money has three functions:

1. measure;
2. exchange;
3. preservation.

However, the following function should also be included:

4. self-growth.

Self-growth is not restricted to growth by interest in a bank account, but also includes investment and even speculation, which sometimes causes a decrease.

For a small amount of money, functions 1, 2, and 3 are dominant, and function 4 is negligible. For an intermediate amount of money, say of the order of \$10 000, all the four functions operate. However, for a huge amount of money, functions 1 and 2 become less relevant and function 4 becomes dominant. Function 2 becomes irrelevant because we cannot buy a hamburger with a 1 billion dollar check, and function 1 is also irrelevant because we cannot easily tell what we can buy with such a huge amount of money. The important point is that function 4 has basic statistical properties.

The target of econophysics is generally not restricted to any special topic, and it seems that understanding function 4 in relation to money is the present goal of this new field of science.

References

Aoyama H, Souma W, Nagahara Y et al. (2000) Pareto's law for income of individuals and debt of bankrupted companies. Fractals 8:293–300

Bak P (1996) How nature works: the science of self-organized criticality. Springer, New York

Bouchaud J-P, Potters M (2000) Theory of financial risks—from statistical physics to risk management. Cambridge University Press, Cambridge

Gopikrishnan P, Meyer M, Amaral L et al. (1998) Cubic law for the distribution of stock price variations. Eur Phys J B3:139–140

Hirabayashi T, Takayasu H, Miura H et al. (1993) The behavior of a threshold model of market price in stock exchange. Fractals 1:29–40

Mandelbrot BB (1997) Fractals and scaling in finance. Springer, Berlin, Heidelberg, New York, pp 371–418

Mantegna RN, Stanley HE (1995) Scaling behavior in the dynamics of an economic index. Nature 376:46–49

Mantegna RN, Stanley HE (2000) An introduction to econophysics. Cambridge University Press, Cambridge, pp 68–75

Okuyama K, Takayasu M, Takayasu H (1999) Zipf's law on income distribution of companies. Physica A 269:125–131

Pareto V (1897) Le cours d'économie politique. Macmillan, London

Sato A-H, Takayasu H, Sawada Y (2000) Power law fluctuation generator based on an analog electrical circuit. Fractals 8:219–225

Takayasu H (1989) Steady-state distribution of a generalized aggregation system with injection. Phys Rev Lett 63:2563–2566

Takayasu H (1990) Fractals in the physical sciences. Manchester University Press, Manchester

Takayasu H, Sato A-H, Takayasu M (1997) Stable infinite variance fluctuations in randomly amplified Langevin systems. Phys Rev Lett 79:966–969

Takayasu H, Takayasu M (1999) Critical fluctuations of demand and supply. Physica A269:24–29

Takayasu H, Takayasu M, Okazaki PM et al. (2000) Fractal properties in economics. In: Novak MM (eds) Paradigms of complexity. World Scientific, Singapore, pp 243–258

The Master Equation Approach to Self-Organization in Labor Mobility

MINORU TABATA*, NOBUOKI ESHIMA†, AND ICHIRO TAKAGI‡

Summary. The master equation is a nonlinear integropartial differential equation that has its origin in statistical physics. The master equation can describe the labor mobility motivated by regional economic disparity in mathematical socioeconomics. By making use of functional–analytical methods and numerical–analytical methods, we deduce that the mathematical economic model described by the master equation is strongly governed by a certain dominant parameter, as follows: if the parameter is smaller than a certain critical positive constant ($\frac{1}{2}$), then the model has an asymptotically stable unique equilibrium, but if the parameter is larger than the constant, then the model has an infinite number of unstable equilibria and exhibits typical self-organization when the time variable is small, but the spatial structure thus self-organized crumbles as the time variable increases.

Key words. Labor mobility, Master equation, Functional–analytical approach, Numerical–analytical approach, Self-organization

1 Introduction

Large free economic unions such as the European Union (EU) and the North American Free Trade Agreement (NAFTA) have recently been established. In such free-trade unions, goods are traded freely, but interregional labor mobility is restricted at a certain level of rigidity. However, there is now a move to abolish this restriction entirely. If no restriction is imposed on regional labor mobility, and if regional economic disparity exists, then workers will move in order to achieve a higher income. This socioeconomic phenomenon describes labor mobility motivated by regional economic disparity, and it has been shown (Helbing 1995; Weidlich and Haag

* Department of Applied Mathematics, Faculty of Engineering, Kobe University, Kobe 657-8501, Japan
† Department of Medical Information Analysis, Faculty of Medicine, Oita Medical University, Oita 879-5593, Japan
‡ Department of Management Engineering, Faculty of Engineering, Kyushu Tokai University, Kumamoto 862-8652, Japan

1983, 1988) that the master equation can describe such a phenomenon. See, for example, Åndersson and Philipov (1982), Dendrinos and Sonis (1990), Haag and Weidlich (1986), Hotelling (1978), and Zhang (1989, 1991) for details of the theory of interregional migration.

The master equation has its origin in statistical physics. The equation is a nonlinear integropartial differential equation, which has the following form:

$$\partial v(t,x)/\partial t = -w(t,x)v(t,x) + \int_{y \in D} W(t;x|y)v(t,y)\mathrm{d}y \tag{1}$$

where D is a bounded domain included in the two-dimensional Euclidean space. By $v = v(t, x)$, we denote the unknown function representing the labor density at time $t \geq 0$ and at a point $x \in D$. By $W = W(t; x|y)$, we denote the transition rate of labor at time $t \geq 0$ and from a point $y \in D$ to a point $x \in D$. The coefficient $w = w(t, x)$ is defined as

$$w = w(t,x) := \int_{y \in D} W(t;y|x)\mathrm{d}y \tag{2}$$

where "$A := B$" denotes that the symbol A on the left-hand side is defined by the expression B on the right-hand side. The first term (the second term) of the right-hand side of Eq. 1 indicates the total amount of labor which goes out of $x \in D$ (comes into $x \in D$) per unit of time.

If we succeed in constructing a mathematical economic model by making use of a nonlinear functional equation such as, e.g., the master equation, then in order to ascertain the validity of the model, we need to investigate fully what kind of economic results can be deductively derived from the model. Even if each assumption of the model seems very plausible, then there is still a fear that some of economic results derived from the model cannot be observed in the real economic world. Hence, without this important investigation, we cannot say that the validity of the model has been demonstrated completely. Furthermore, we have to study the behavior of the model in order to investigate the real economic world by making use of the model. Therefore, in this chapter we study in detail what kind of economic results can be deductively derived from the mathematical economic model described by the master equation.

However, in general, the behavior of solutions of nonlinear functional equations is extremely complicated. In fact, it is almost impossible to investigate the behavior of solutions of the master equation fully by making use of classical analysis alone. The reason for the difficulty thus encountered is the strong nonlinearity of the master equation. Indeed numerical–analytical approaches have been taken in various studies on nonlinear economic models in order to observe the complicated behavior of these models. However, the strong nonlinearity of the master equation makes the model behave in an extremely complicated manner, and hence, we cannot observe the behavior of the model accurately by numerical–analytical methods alone. Therefore, we need to employ more advanced mathematical methods than numerical–analytical methods. Recalling that the master equation has its origin in statistical physics, and noting that various functional–analytical methods are usefully employed in statistical physics, we find it useful to take a functional–analytical approach to the model described by the master equation. In this chapter, we employ both numerical–analytical methods and functional–analytical methods. The former approach gives us a plain picture of the behavior of the model, and the latter gives us an accurate picture of the behavior of the model.

2 The Mathematical Economic Model

We consider the initial value problem for the master equation. If a function $\nu = \nu(t, x)$ satisfies Eq. 1 for each $t > 0$ and the following initial condition at $t = 0$:

$$\nu(0, x) = \nu_0(x) \tag{3}$$

where $\nu_0 = \nu_0(x)$ is a given positive-valued function, then we say that the function $\nu = \nu(t, x)$ is a *solution* to the initial value problem with the initial condition Eq. 3. It has been proved at the mathematical level of rigor that the initial value problem has a unique positive-value solution (see Tabata et al. (1998, 1999) and Tabata and Eshima (2000) for details). First we make the assumption described below.

Assumption 2.1: If the labor density is equal to $\nu_0 = \nu_0(x)$ at time $t = 0$, then the labor density at time $t > 0$ is equal to a solution $\nu = \nu(t, x)$ of the initial value problem with the initial condition in Eq. 3.

The above assumption is obtained as a result in Helbing (1995) and Weidlich and Haag (1988), but here we regard it as an assumption in the same way as in Tabata et al. (1998, 1999) and Tabata and Eshima (2000). We will restrict the form of the transition rate $W = W(t; x|y)$ as described below.

Assumption 2.2: The transition rate $W = W(t; x|y)$ has the form

$$W(t; x|y) = \alpha(t) \exp\{U(t,x) - U(t,y) - E(x,y)\}$$

where $\alpha = \alpha(t)$ denotes the flexibility at time $t \geq 0$, $U = U(t, x)$ is the utility at time $t \geq 0$ and at a point $x \in D$, and $E = E(x, y)$ denotes the effort from a point $y \in D$ to a point $x \in D$.

Assumption 2.2 is also employed by Helbing (1995 pp. 137–138), Weidlich and Haag (1988 pp. 81–100), and Tabata and Eshima (2000). We will discuss the flexibility, utility, and effort (see Helbing (1995 pp. 137–157) for these quantities). As fully investigated by Weidlich and Haag (1988 pp. 81–100), various socioeconomic and economic variables have an influence on the utility. If we choose socioeconomic and economic variables appropriately, then we can regard the utility as a certain function of the chosen variables. For example, it is known that the utility is strongly affected by the following socioeconomic and economic variables: (1) unemployment rate, (2) export structure index (industrial exports minus agricultural exports divided by total exports), (3) over-night stays per capita, (4) percentage of total employment in the tertiary sector, and (5) labor density. The variables 1–5 were chosen by Weidlich and Haag (1988 pp. 81–100), and are called *key factors*.

However, even if we choose the key factors listed above, the evolution of the regional labor mobility is still extremely complicated. Hence, in order to study these phenomena with the emphasis on an investigation into the fundamental mechanism driving them, we build a mathematical economic model on the basis of assumptions which are as simple as possible, i.e., we will follow *the KISS principle* (keep it simple, stupid) which is advocated by R. Axelrod. Moreover, it was shown by Weidlich and Haag (1988) that the utility depends mainly upon key factor 5 (see Weidlich and Haag (1988 Sect. 4.4.2)). From these considerations, we now choose only the labor density as a key factor, and we impose the following assumption on the utility. This was also employed by Tabata et al. (1999) and Tabata and Eshima (2000).

Assumption 2.3: The utility $U = U(t, x)$ is an affine function of the ratio $v(t, x)/\|v(t, \cdot)\|_{L^1(D)}$ of the form

$$U(t,x) = c_{2.1}v(t,x)/\|v(t,\cdot)\|_{L^1(D)} + c_{2.2}$$

where $c_{2,j}$, $j = 1, 2$, are positive constants, and we define the norm $\|v(t, \cdot)\|_{L^1(D)}$ as

$$\|v(t,\cdot)\|_{L^1(D)} := \int_{y\in D} |v(t,y)| dy \tag{4}$$

Remark 2.4:

(i) By $L^1(D)$ we denote the function space of all real-valued Lebesgue integrable functions defined in D (see, e.g., Yosida (1980) for details of $L^1(D)$).
(ii) We can easily see that $\|v(t, \cdot)\|_{L^1(D)}$ denotes the total amount of labor at time $t \geq 0$. By integrating both sides of Eq. 1 with respect to $x \in D$ in the same way as was done by Tabata et al. (1998, 1999) and Tabata and Eshima (2000), we can easily obtain a conservation law for the total amount of labor, i.e.,

$$\|v(t,\cdot)\|_{L^1(D)} = \|v_0(\cdot)\|_{L^1(D)} \qquad \text{for each} \qquad t \geq 0 \tag{5}$$

Hence we can rewrite $U = U(t, x)$ as

$$U(t,x) = c_{2.1}v(t,x)/\|v_0(\cdot)\|_{L^1(D)} + c_{2.2} \tag{6}$$

It follows from assumption 2.3 that the utility is a monotonously increasing function of the ratio

$$q := v(t,x)/\|v_0(\cdot)\|_{L^1(D)} \tag{7}$$

If the utility $U = U(t, x)$ is a more complicated increasing function of the ratio q than Eq. 6, then we can regard the affine function in Eq. 6 as a linear approximation for the increasing function. It follows from assumption 2.3 that only *imitative processes* and no *avoidance processes* contribute to labor mobility (see Helbing (1995) for these processes).

We now discuss flexibility. It was shown by Weidlich and Haag (1988 pp. 81–100) that the following additional socioeconomic and economic variables have a great influence on flexibility: (6) total labor, (7) total job vacancies, (8) investment structure index (expansionary minus rationalizing investment divided by total investment), and (9) total income (total production). Variables 6–9 were chosen as key factors by Weidlich and Haag (1988 pp. 81–100). Neglecting key factors 7–9 for the same reason as given in assumption 2.3, and applying Eq. 5, we now impose an assumption on the flexibility.

Assumption 2.5: The flexibility is identically equal to a positive constant.

We now discuss the effort. It is plausible to assume that the effort $E = E(x, y)$ is a monotonously increasing function of cost incurred in moving from a point $y \in D$ to a point $x \in D$. Furthermore, in general, the cost depends on the distance $|x - y|$. However, if D is not very large, then the cost is independent of $|x - y|$. Hence, for simplicity, we reasonably assume that if $x \neq y$, then the cost is identically equal to a positive constant. Therefore we now make a further assumption.

Assumption 2.6: If $x \neq y$, then the effort from a point $y \in D$ to a point $x \in D$ is identically equal to a positive constant.

3 Equilibria of the Model

If the master equation has a solution that is independent of the time variable, then we call that solution a *stationary solution*. Each equilibrium is expressed by a stationary solution. We now classify the equilibria of the model. Suppose that each *neighborhood* of an equilibrium contains a function of the space variable such that if the initial function is equal to this function, then the initial value problem has a solution that moves away from the equilibrium infinitely as the time variable tends to infinity. We then say that the equilibrium is *unstable*. Suppose that an equilibrium has a neighborhood such that if the initial function is contained in that neighborhood, then the initial value problem has a solution that converges to the equilibrium as the time variable tends to infinity. We then say that the equilibrium is *asymptotically stable.*

Remark 3.1: It follows from the definition of equilibrium that each equilibrium of the model is expressed not by a point of a finite-dimensional Euclidean space, but by a function of the space variable. Hence, we need to define the words "neighborhood," "converge," and "move away" not in Euclidean spaces, but in a function space defined in D. Here, we employ the function space $L^\infty(D)$, where $L^\infty(D)$ denotes the set of all essentially bounded, real-valued functions defined in the domain D (see, e.g., Yosida (1980) for details of $L^\infty(D)$).

The equilibria of the model greatly depend on whether the parameter $c_{2.1}/|D|$ is smaller than 1/2 or not, where $c_{2.1}$ is the constant defined in assumption 2.3, and $|D|$ denotes the Lebesgue measure of the domain D. We now deal with the case where $c_{2.1}/|D| < 1/2$. Making use of functional–analytical methods, we can obtain estimates of solutions of the master equation given by Tabata et al. (1998, 1999) and Tabata and Eshima (2000). Applying these results to the model constructed in Sect. 2, we obtain the theorem given below.

Theorem 3.2: If $c_{2.1}/|D| < 1/2$, then the model has a unique asymptotically stable equilibrium, which is expressed by the positive constant $\|v_0(\cdot)\|_{L^1(D)}/|D|$.

See Eq. 3 for $v_0 = v_0(x)$. We now deal with the case where $c_{2.1}/|D| > 1/2$. If a function of the space variable is constant in a subset d and in its complement $D\backslash d$, then we say that the step function is a *two-step function.* Making use of functional–analytical methods, we can obtain estimates of solutions of the master equation given by Tabata et al. (1998, 1999) and Tabata and Eshima (2000). Applying these results to the model constructed in Sect. 2, we can obtain the theorem given below.

Theorem 3.3: If $c_{2.1}/|D| > 1/2$, then the model has an infinite number of unstable equilibria, which are expressed by positive-valued, two-step functions.

The details of the proof of theorems 3.2–3 are given by Tabata and Eshima (2000). We cannot deal with the case where $c_{2.1}/|D| = 1/2$ by making use of our method. We now discuss theorems 3.2–3. From theorems 3.2–3, we see that the behavior of the model greatly depends on whether the parameter $c_{2.1}/|D|$ is smaller than 1/2 or not. Thus, we can say that $c_{2.1}/|D|$ is a dominant parameter which strongly governs the model, and that the constant 1/2 is its critical value. From Eq. 6, we can consider that the coefficient $c_{2.1}$ indicates the strength of the dependence of the utility upon the ratio q (see Eq. 7). Broadly speaking, we can say that the Lebesgue measure of D indicates the size of D. Hence, we can say that if the domain D is large and if the utility does not greatly depend on the ratio q, then the model has an asymptotically stable unique equilibrium. Conversely, if the domain D is small and if the utility depends strongly on q, then the model has an infinite number of unstable equilibria.

It is almost impossible to remove small error completely in statistical measurements. Therefore, even if the initial labor density is equal to an equilibrium at the statistical level of rigor, then there exists a very small gap between the initial labor density and the equilibrium in the real economic world. Furthermore, the model will encounter unpredictable small fluctuations as the time variable increases, and there is the possibility that such fluctuation will make the gap larger and larger. If the equilibrium is asymptotically stable, then the gap cannot become large, and it will converge to 0 as the time variable tends to infinity. Small errors and small fluctuations have no important influence on the behavior of the model. However, if the equilibrium is unstable, then the behavior of the model is extremely complicated. The labor density cannot remain in the neighborhood of the unstable equilibrium, and there is the possibility that the labor density will move away from the unstable equilibrium as the time variable increases. Even extremely small unpredictable fluctuations can have a large influence on the behavior of the labor density. As a result of these considerations, we need to investigate fully how the labor density moves away from the unstable equilibria as the time variable increases. We now attempt to take a numerical–analytical approach to this subject in the next section.

4 Numerical–Analytical Approach

For simplicity, we assume that D is a square such that $D := [0, a) \times [0, a)$, where a is a positive constant. It follows from assumptions 2.1–3, (Eq. 6), and assumptions 2.5–6 that $W = W(t; x|y)$ has the form

$$W(t;x|y) = c_{4.1} \exp\{c_{2.1}(v(t,x) - v(t,y))/\| v_0(\cdot) \|_{L^1 D}\} \tag{8}$$

where $c_{4.1}$ is a positive constant. We now rewrite Eq. 1 with Eq. 8 by introducing the new unknown function $u = u(t, x)$ in place of $v = v(t, x)$:

$$u = u(t,x) := c_{2.1} v(t/c_{4.1}a^2, ax)/\|v_0(\cdot)\|_{L^1(D)} \tag{9}$$

In exactly the same way as given by Tabata et al. (1999) and Tabata and Eshima (2000), we deduce that $u = u(t,x)$ satisfies the integropartial differential equation of the simple form

$$\partial u(t,x)/\partial t = -u(t,x)e^{-u(t,x)} \int_{y\in\Omega} e^{u(t,y)} dy + e^{u(t,x)} \int_{y\in\Omega} u(t,y) e^{-u(t,y)} dy \tag{10}$$

where $\Omega := [0,1) \times [0,1)$. From Eq. 9, we can easily see that

$$\int_{y\in\Omega} |u(0,y)| dy = c_{2.1}/a^2, \qquad |D| = a^2 \tag{11}$$

We divide the domain of the time variable $[0, +\infty)$ as follows:

$$[0,+\infty) = \bigcup_{i\in\mathbb{N}} [(i-1)\Delta\tau, i\Delta\tau)$$

where $\Delta\tau$ is a sufficiently small positive constant. We divide the domain Ω into disjoint small squares as follows:

$$\Omega = \bigcup_{j,k=1,\ldots,N} [(j-1)/N, j/N) \times [(k-1)/N, k/N)$$

where N is a sufficiently large positive integer. By regarding the unknown function $u = u(t, x)$ as constant in each subset of $[0,+\infty) \times \Omega$ of the form

$$[(i-1)\Delta\tau, i\Delta\tau) \times [(j-1)/N, j/N) \times [(k-1)/N, k/N) \tag{12}$$

we can derive the following difference scheme from Eq. 10:

$$\begin{aligned} &\{U(i+1;j,k) - U(i;j,k)\}/\Delta\tau \\ &= -U(i;j,k)\{\exp(-U(i;j,k))\}\sum_{m,n=1,\ldots,N} (1/N^2)\exp(U(i;m,n)) \\ &\quad + \{\exp(U(i;j,k))\}\sum_{m,n=1,\ldots,N} (1/N^2) U(i;m,n)\exp(-U(i;m,n)) \end{aligned} \tag{13}$$

Making use of the difference scheme Eq. 13, we can construct numerical solutions to the initial value problem for Eq. 10.

We now discuss a numerical solution of Eq. 10. Substitute $\Delta\tau = 10^{-9}$ and $N = 128$ into Eq. 13. We assume that $u = u(0, x)$ is very close to 1 but is not identically equal to 1. Hence $\int_{y\in\Omega} |u(0,y)| dy$ is nearly equal to 1. Therefore, from Eq. 11 we see that $c_{2.1}/|D|$ is nearly equal to 1. Hence we can apply theorem 3.3. The graphs of $u = u(t, x)$ are given in Figs. 1–20. Inspecting these figures, we see that the evolution of the labor density has two different stages. Figures 1–10 and Figs. 11–20 describe the first and second stage, respectively. The numerical solution exhibits typical self-organization in the first stage. However, the spatial structure thus self-organized in the first stage crumbles in the second stage in such a complicated way that $u = u(t, x)$ tends to infinity in a certain subset d, but converges to 0 in its complement $D \backslash d$. Inspecting Figs.

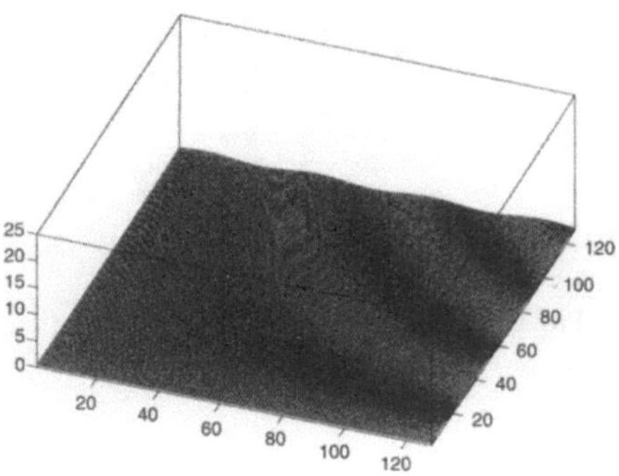

FIG. 1. $t = 0$

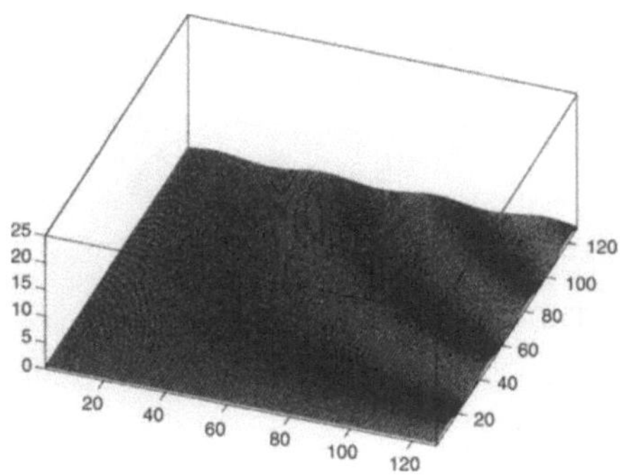

FIG. 2. $t = 0.125$

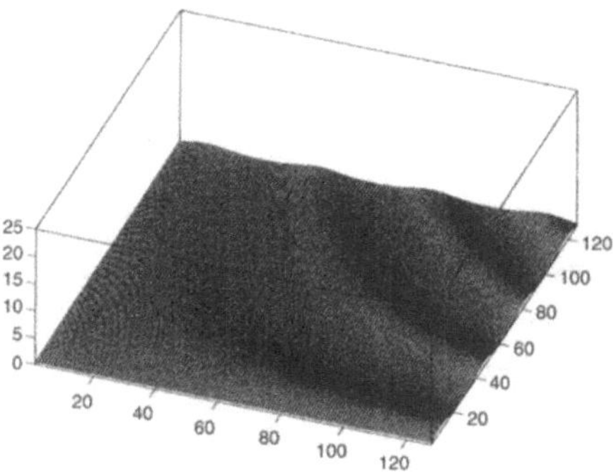

FIG. 3. $t = 0.625$

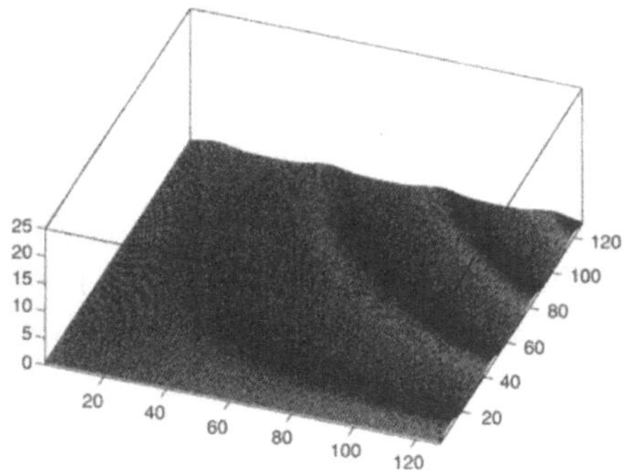

FIG. 4. $t = 1.25$

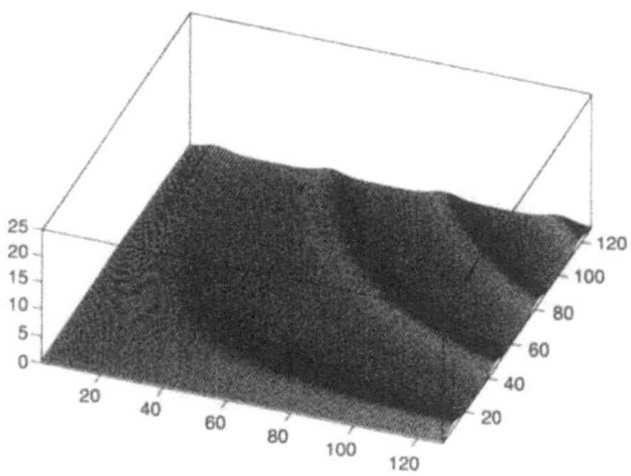

FIG. 5. $t = 1.875$

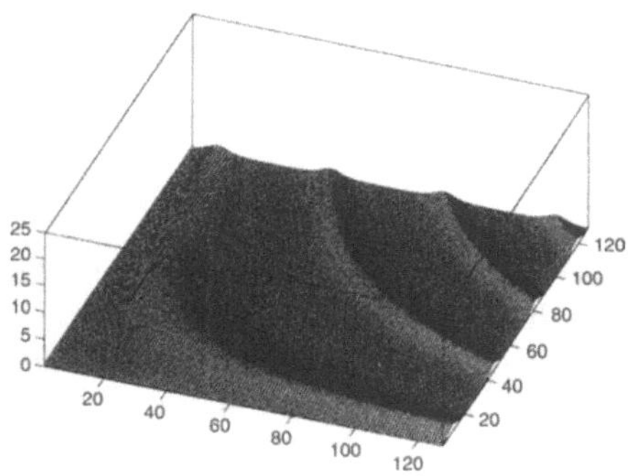

FIG. 6. $t = 2.5$

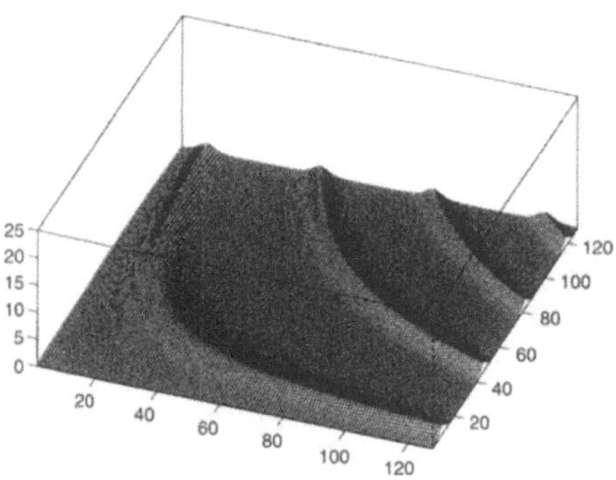

FIG. 7. $t = 3$

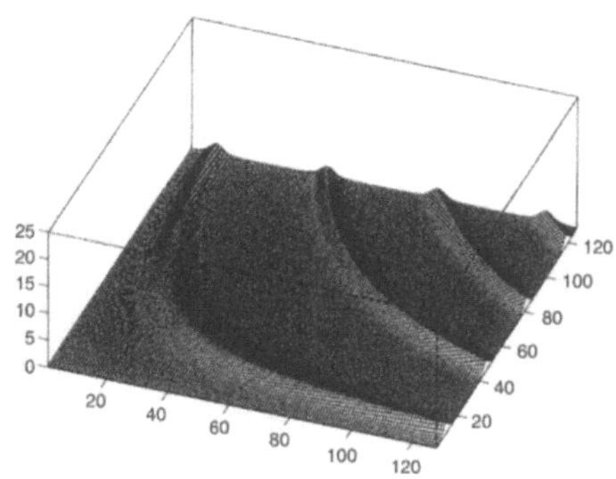

FIG. 8. $t = 3.25$

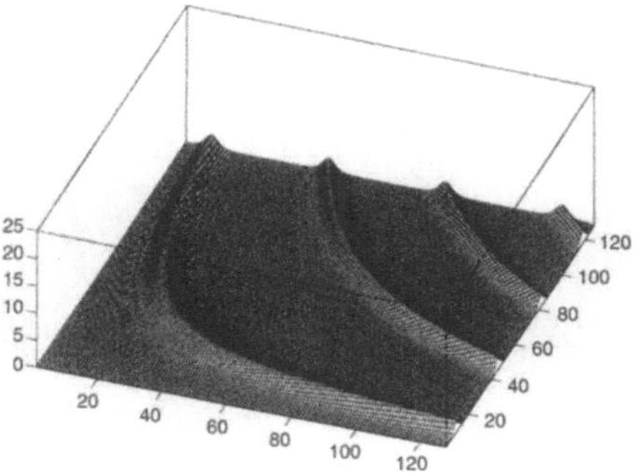

FIG. 9. $t = 3.5$

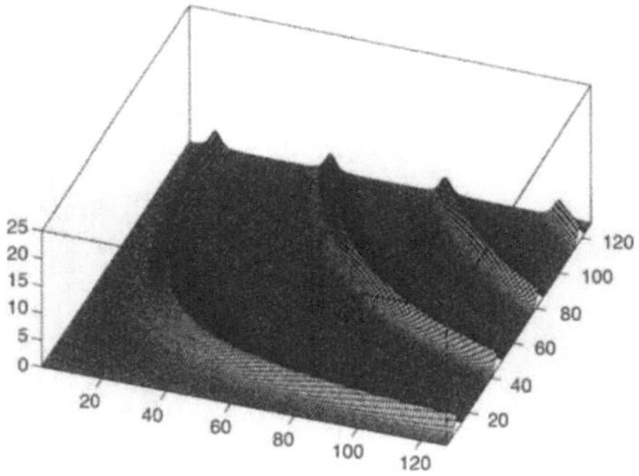

FIG. 10. $t = 3.75$

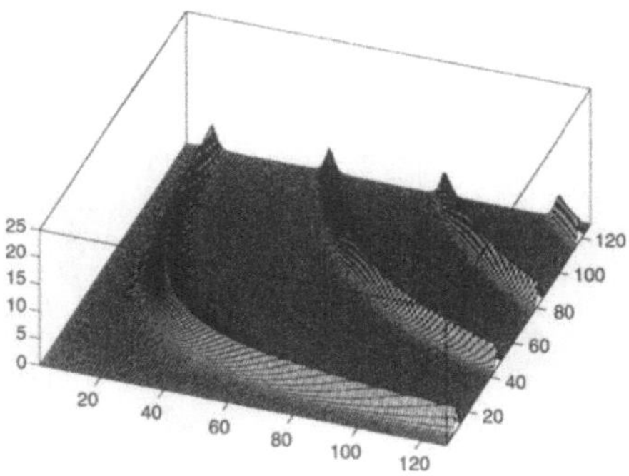

FIG. 11. $t = 4$

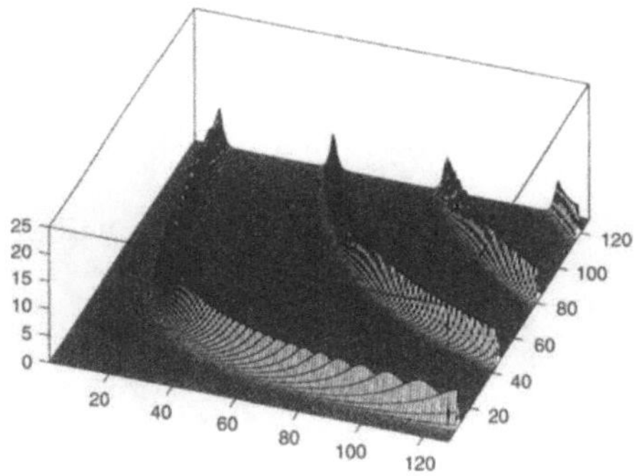

FIG. 12. $t = 4.140625$

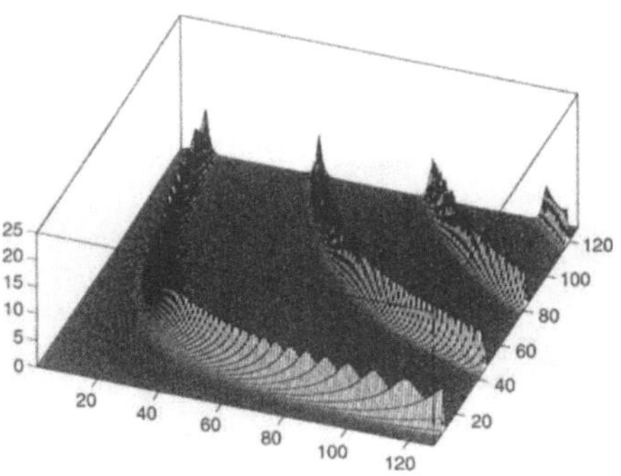

FIG. 13. $t = 4.156250$

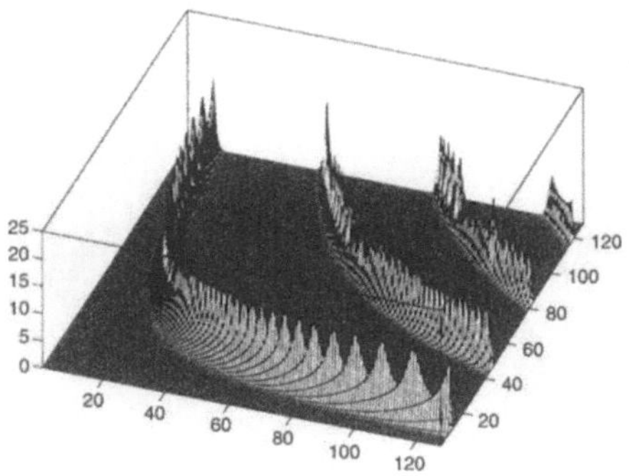

FIG. 14. $t = 4.171875$

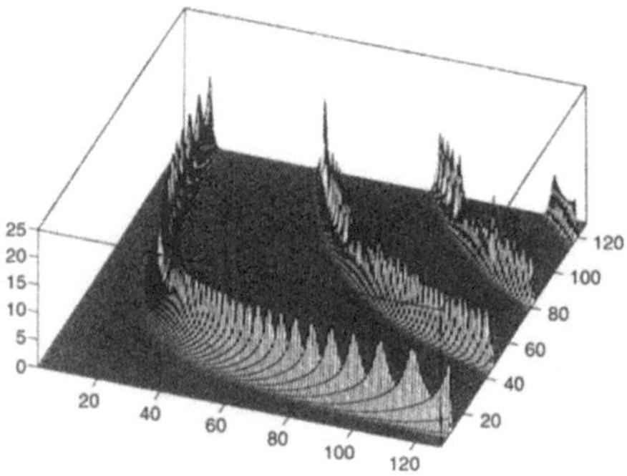

FIG. 15. $t = 4.171880722045898$

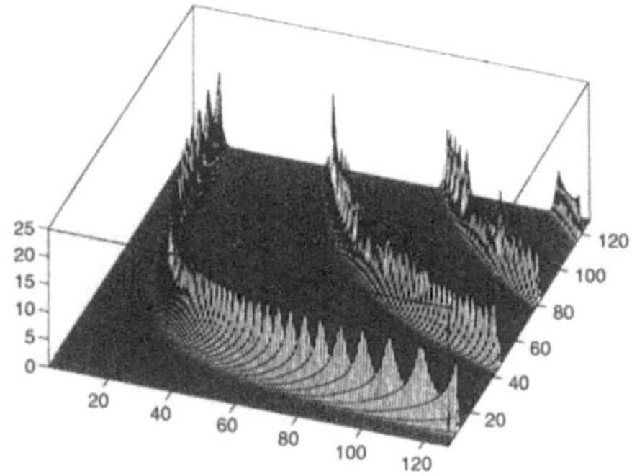

FIG. 16. $t = 4.171886444091797$

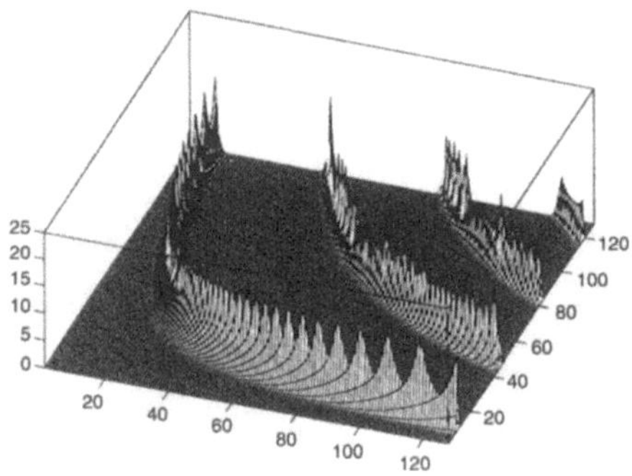

FIG. 17. $t = 4.171890258789062$

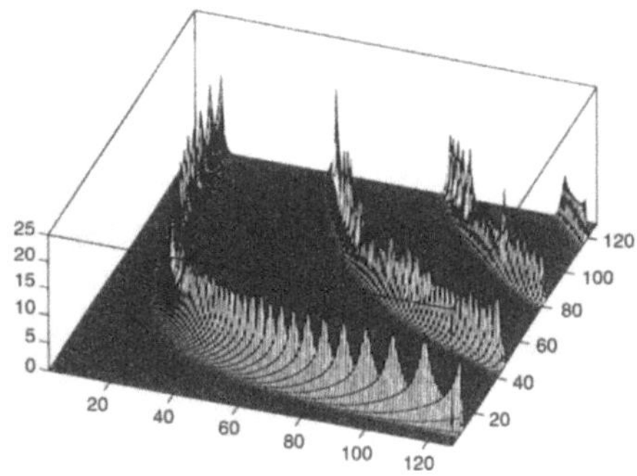

FIG. 18. $t = 4.171897888183594$

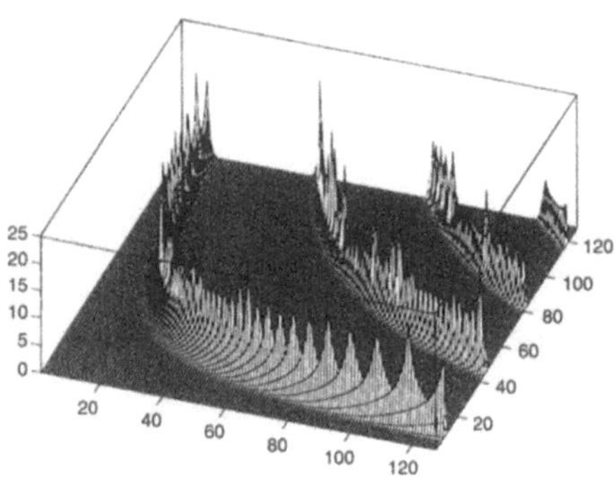

FIG. 19. $t = 4.171900749206543$

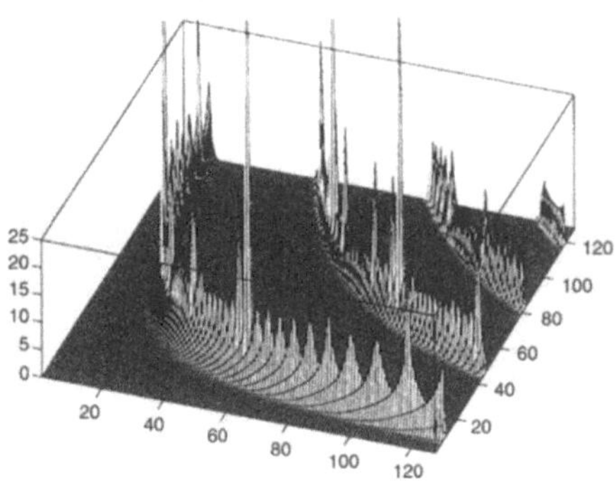

FIG. 20. $t = 4.171901701702880859$

11–20, we see that the behavior of $u = u(t, x)$ is extremely complicated. We find it almost impossible to observe the behavior of the labor density fully as the time variable increases by employing only the numerical–analytical method. Hence, by making use of functional–analytical methods, we investigate the labor density more fully in the next section.

5 The Over-Concentration of Labor

Applying functional–analytical results obtained by Tabata et al. (1998, 1999) and Tabata and Eshima (2000) to the model constructed in Sect. 2, we obtain the theorem given below.

Theorem 5.1: If $c_{2.1}/|D| > 1/2$, and if the Lebesgue measure of $\{x \in D;\ v_0(x) = \text{ess sup}_{x\in D}\, v_0(x)\}$ is equal to 0, then a subset $d(r){:} = \{x \in D;\ v(r, x) \geq \|v_0(\cdot)\|_{L^1(D)}/|D|\}$ satisfies (i–iii) below, where $r \geq 0$ is the parameter.

(i) The Lebesgue measure of $d(r)$ is not equal to 0 for each $r \geq 0$, and converges to 0 as $r \to +\infty$.
(ii) If $r \geq s$, then $d(r) \subseteq d(s)$.
(iii) $\lim_{t\to+\infty} \text{ess sup}_{x\in D\setminus d(r)}\, v(t, x) = 0$, $\lim_{t\to+\infty} \text{ess sup}_{x\in d(r)}\, v(t, x) = +\infty$, and $\lim_{t\to+\infty} \text{ess inf}_{x\in d(r)}\, v(t, x) = 0$, for each $r \geq 0$.

Tabata and Eshima (2000) give the proof of this theorem. By making use of theorem 5.1, we can fully explain the behavior of the labor density. We can easily see that the subset $d(r)$ corresponds to the subset d defined in Sect. 4.

It follows from theorem 5.1 that the Lebesgue measure of $d(r)$ becomes smaller and smaller and converges to 0 as r tends to infinity. Hence, we see that the Lebesgue measure of $\cap_{r\geq 0} d(r)$ is equal to 0. However, making use of Eq. 5 and theorem 5.1, (iii), we can say that all the labor moves into $\cap_{r\geq 0} d(r)$. Hence, all the labor is concentrated in a subset whose Lebesgue measure is equal to 0, i.e., over-concentration of labor is observed.

By theorem 5.1, (iii), we see that

$$\text{ess sup}_{x\in d(r)}\, v(t,x) - \text{ess inf}_{x\in d(r)}\, v(t,x) \to +\infty \text{ as } t \to +\infty$$

for each $r \geq 0$. It follows from this result that the labor density behaves in $d(r)$ as if its graph were infinitely pulled in the vertical direction. We can reasonably consider that the vertical virtual force acting upon the graph and the over-concentration of labor mentioned above make the spatial structure in the first stage (see Figs. 1–10) crumble in the second stage (see Figs. 11–20) as the time variable increases.

Acknowledgment. The authors would like to express their deepest gratitude to Dr. Y. Takei for performing the numerical calculations.

References

Åndersson AE, Philipov D (1982) Economic models of migration. In: Albegov M, Åndersson AE, Snickars F (eds) Regional development modeling: theory and practice. North Holland, Amsterdam
Dendrinos DS, Sonis M (1990) Chaos and socio-spatial dynamics. Springer, New York
Haag G, Weidlich W (1986) A dynamic migration theory and its evaluation of concrete systems. Reg Sci Urban Econ 16
Helbing D (1995) Quantitative sociodynamics. Kluwer Academic, Dordrecht
Hotelling H (1978) A mathematical theory of migration. Environ Plann A10:1225–1239
Tabata M, Eshima N (2000) The behavior of solutions to the Cauchy problem for the master equation. Appl Math Comput 112:79–98
Tabata M, Eshima N, Takagi I et al. (1998) The Cauchy problem for the system of equations describing migration motivated by regional economic disparity. Appl Math Comput 94:45–64
Tabata M, Eshima N, Takagi I (1999) The nonlinear integro-partial differential equation describing the logistic growth of human population with migration. Appl Math Comput 98:169–183
Weidlich W, Haag G (1983) Concepts and models of a quantitative sociology. Springer, Berlin
Weidlich W, Haag G (eds) (1988) Interregional migration. Springer, Berlin
Yosida K (1980) Functional analysis, 6th edn. Springer, Berlin, Heidelberg, New York
Zhang WB (1989) Coexistence and separation of two residential groups—an interactional spatial dynamic approach. Geogr Anal 21:91–102
Zhang WB (1991) Time and change in nonlinear economics. Springer, Berlin

Part VI Economic Heresies

From "Historical Time" to the Economics of Complexity

Masaaki Yoshida

Summary. I first introduce a new interpretation of Keynes' *Treatise on Money*, which shows the way to construct an economic theory based on irreversible time. Second, I show the more general framework as an interactive network system composed of satisficing agents working in irreversible time. This framework makes it possible to analyze our economy as a real working system.

Key words. Keynes, Historical time, Satisficing principle, Neural network, Buffer

1 Introduction

Robinson (1971), in the preface to the Japanese edition of 1973, emphasized that the point of the Keynesian revolution is in his clear recognition of the irreversibility of economic decisions and behavior. Some post-Keynesian economists (e.g., Kregel 1975; Eichner 1979; and others) also pursued this point, and discussed the importance of constructing an economic theory on the basis of "historical time" (i.e., irreversible time). However, the post-Keynesians confine themselves to emphasizing the role of uncertainty in the actual economy, and do not attempt to construct an economic theory on the basis of historical time. On the other hand, historians studying the development of the "general theory" also have not shown how and when Keynes actually formulated economics based on historical time.

This chapter introduces some Japanese results on this theme to English readers, and I will concentrate on two points in particular.

1. From a study of the *Treatise on Money* (Keynes 1971), we can easily get the basic design of a new economics based on historical time.
2. On the basis of this design, we can construct an economic theory as an interactive network composed of satisficing agents operating under irreversible time.

Before that, in the next section, we will clarify the reason why irreversible time is important in an analysis of our economy as a real working system.

Faculty of Economics, Senshu University, Higashi-Mita 2-1-1, Tama-ku, Kawasaki 214-8580, Japan

2 Some Problems with the Basic Architecture of Economics

Neoclassical economics has no view about how economic models can operate as real working systems. If you want to discuss how a system can work in real time, you cannot do so without considering what kind of real time constraints agents face, and how they deal with problems in the real world. It is necessary to make clear, in a bottom-up manner, what kind of mechanisms let the system work. However, this point has not been taken seriously. I think this is because neoclassical economics has a rather strange structure concerning time (Fig. 1a).

In the general equilibrium theory, it is assumed that any calculations for optimization and any contracts relating to it have no effect on agents. In the general equilibrium theory, agents are introduced as machines which answer questions about demand or supply according to the price information. However, if the ordinal size of commodities is taken into consideration, it is impossible to complete this task even with a supercomputer. This is the problem of an explosion in calculations, which has repeatedly been reported by Yoshinori Shiozawa (see Shiozawa 1990). It is clear that there is a lack of attention to what occurs in real working systems. Neoclassical economics simply assumes that agents can do complex calculations. However, things are even worse than this. The results of such onerous tasks become nothing if all demands and supplies coincide, and the optimizing calculations are repeated under new price vectors until an equilibrium is reached. Meanwhile, there are no changes to the initial endowments, utility functions, and variable circumstances of the agents. The time

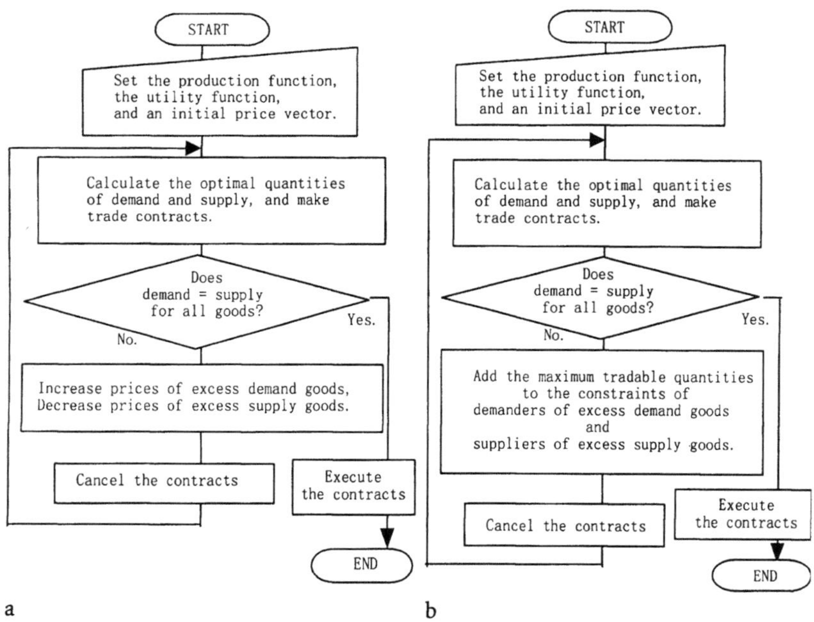

FIG. 1. The reversible time structure of neoclassical economics

structure is completely crushed because from the point of view of the agents, who should be treated as the foundation for the mechanisms of the system, no action has any effect.

Things are not improved even in the case of the general nonequilibrium model, which approves tradeing under disequilibrium prices. It is often said that agents in this model actually engage in economic activity, exchange information, and adjust their behavior (e.g., Negishi 1980 p. 80), but, as can be seen in Fig. 1b, although the information supplied shifts from prices to tradable quantities, the resulting economic activity is assumed not to benefit the agents either.

There are other justifications for attention being concentrated on optimized equilibrium. It is said that things will converge to an equilibrium state after the agents have faced the same situation repeatedly, even if time flows irreversibly for those agents. Agents with learning ability are thought to reach an equilibrium state both in the rational expectation macroeconomic model which assumes the perfect rationality of agents, and in evolutionary game theory which excludes the optimizing behavior of agents. However, there is no reason why the situation which agents face should remain the same, so there is also no consistency between the behavior of agents and the behavior of the system.

With such a strange time structure, the problem of an explosion of calculations does not matter. The uncertainty under irreversible time is excluded by this setting of the time structure. Things which might occur in nonequilibrium states are regarded as irrational and are expected to disappear quickly, so they are ignored in mainstream economics.

However, if it seems necessary to have a theoretical system which facilitates comprehension of the real working economic system, it is no longer useful to persist with neoclassical economic theories. Therefore, in the following sections we propose a basic architecture for a theoretical model which can operate in real time. The clue can be found in Keynes' *Treatise on Money*.

3 A New Interpretation of the *Treatise on Money*

In most of previous studies of the development of the "general theory" (GT), the *Treatise on Money* (TM) has been treated as just a vehicle to carry the GT. It has commonly been accepted that the TM is written in traditional terms which contrast with the revolutionary GT, and many historians have tried to follow a logical progression from the TM to the GT. However, this popular story has two serious defects.

1. In what way is the GT theoretically revolutionary? Although 50 years have passed since Keynes' death, it becomes more and more difficult to get agreement among economists on the theoretical structure of the GT. Therefore, this version of the story does not achieve its goal.
2. What constitutes the traditional world where the TM belongs? A careful reading of the book will reveal the differences between the TM and the classics, so the story also loses its starting point.

Thus, today, all historians of Keynesian economics must develop their own theoretical view based on their own historical study. After the hundreds of Keynesian models which have been made by theorists, you may think that this is almost impossible. However, if you trace the development of Keynes' thoughts, without any theoretical prejudice, from the *Tract on Money* to the GT, you will find that it is not too difficult to find a consistent flow, and to fit it into a consistent framework. We have no space to introduce all the results of all the historical investigations (see Yoshida 1997 Part I), so we will briefly describe the new way to read the TM.

On the fundamental equations, Keynes says, "Their only point is to analyse and arrange our material in what will turn out to be a useful way for tracing cause and effect, when we have vitalised them by the introduction of extraneous facts from the actual world." (Keynes 1971 p. 125). Hence, the key to a clear understanding of the theoretical structure of the TM is to take the fundamental equations as being genuinely fundamental.

As Keynes says, the first fundamental equation is a mere identity, because it just insists that

> Current expenditure on consumption goods = Cost of production of the current output of consumption goods (including the normal remuneration of entre-preneurs) + "Profit" (defined as the sales minus the cost)

The consumption expenditure can be expressed as the money income minus savings, which is defined as the difference between the income and the consumption expenditure, and the money income is defined as the cost of production of consumption goods and investment goods. Then the "profit" can be rewritten as the cost of production of investment goods minus savings. After dividing both side of this equation by the volume of the output of consumption goods, we get the form of the first fundamental equation.

The volume of the output of consumption goods is the result of production activity in the past. So, in this meaning only, it cannot be changed in the current period. Of course, it can be changed in the next period. It is by no means fixed at the full employment level. The price of consumption goods is defined as the sales divided by the volume of the output. So it is simply the average price. This does not imply that all consumption goods are purchased at that price, or that the price is an equilibrium price.

To "vitalize" this equation, we must pay attention to the two group-behavior assumptions which appear in the TM. One is about the output-adjusting behavior of the entrepreneur group which produced the consumption goods. As Keynes says, "the most usual and important occasion of change will be the action of the entrepreneurs, under the influence of the actual enjoyment of positive or negative profits, in increasing or diminishing the volume of employment" (Keynes 1971 p. 142), i.e., the entrepreneurs adjust their output level according to their "profit." This assumption of group behavior is repeatedly mentioned by Toshiaki Hirai as the "TM supply function" (see Hirai 1987). The other assumption is about the consumption expenditure behavior of households according to their income. This assumption is used by Keynes (1971) in Chap. 20.

On the one hand, the adjusting of output by households directly affects their expenditure on consumption goods, so it also affects the sales of the entrepreneurs. On the other hand, the adjusting of output by the entrepreneurs affects the employment level, so it also affects the income level of the households. Thus, putting them together in the fundamental equation, we can easily get the group-behavior foundation of the multiplier process. We showed the formulation and simulation of this process in Yoshida (1988) and Yoshida (1997 pp. 63–66).

Here we must note the following points.

1. Regardless of whether the system is in equilibrium or not, each group actually makes its decisions and behaves autonomously in irreversible time.
2. Each group makes its adjustment decisions by referring to "normal" concepts. This is not optimization behavior, and we prefer to call it satisficing behavior.

In the TM, we see more examples of group behavior (i.e., of entrepreneurs in the investment sector, asset holders, and the banking system), and Keynes' discussion is founded on simulations of the interactions among those group behaviors. To understand such a work, the general equilibrium theory is quite inappropriate. We must have an interactive network system and an open nonequilibrium system in our minds.

In the next section, we show the formulation of a more general system working in irreversible time.

4 The Basic Architecture for the Economics of Real Working Systems

4.1 The Fundamental Components are Satisficing Agents

First, as the fundamental components of our economic society are human, we must consider the fact that their processing ability is limited. In addition, as our economy is composed of many components acting autonomously, we must consider that the time and timings for information processing of effective behaviors are strictly limited. Hence, we should assume that our components face a real-time constraint on their behavior. Therefore, each segment of the information process should be small, so that the processing can be reduced to a small numbers of comparisons. The range of the input information should also be limited within a small area so that good timing is possible. However, it is not possible to assume a behavior model which will maximize the expected utility.

In relation to this point, we assume that agents follow the satisficing principle as the fundamental components of the economic system. These agents perform the routine work of adjusting behavior as follows: in the case of entrepreneurs, they adjust their production level up or down according to their gain or loss, which is sales minus the satisficing standard. The satisficing standard is defined as the production cost plus a "normal" margin. In the case of shopkeepers, when their stock level reaches the "normal" lower level, they buy to being the stock level to the normal upper level. In the case of householders, they adjust their consumption expenditure level higher or lower according to their salary after tax compared with the salary they "normally" expect, and so on.

In our model, we assume that the starting values of the satisficing standards, or the "normal" levels, are given to each agent from some outside organization. This assumption reflects the fact that agents in our economy belong to some category of business which knows what is normal in that category. Agents cannot appear in any business society without any preconceived ideas, as if they were babies. They joins their business society after some initiation processes. The way in which agents change their satisficing standard will be discussed when we treat the problem of the micro–macro loop (see Sect. 4.7).

4.2 The Economic System is an Interactive Network Composed of Satisficing Agents

Next, we think of a single-layered, mutually connected neural network composed of many such agents. Their connection weights show how each agent attaches importance to the behavior of the other agents when making a decision. As mentioned above, the input information area should be narrow, so almost all connection weight values are set at 0. Of course, the threshold value of each unit corresponds to the satisficing standard.

We then get our fundamental equation as follows:

$$X_j(t+1)=a_j\cdot f_j\left(\sum_i W_{ji}X_j(t)-S_j+Z_j(t)\right)$$

where $x_j(t)$ is the adjusting output of the j-th agent at period t, w_{ji} is the weight needed to estimate the adjustment to the output of the i-th agent as the input for the j-th agent, s_j is the satisficing standard of the j-th agent, $z_j(t)$ is the stimulation to the j-th agent from outside of the system, a_j is the parameter of the maximum possible adjustment, and $f_j(\cdot)$ is an increasing function whose minimum/maximum value is ± 1 and whose output is 0 when the absolute of the input does not exceed b_j. Here, b_j reflects the buffer in each information process (see Sect. 4.4).

4.3 The Features of this System

We have briefly shown the outline of an economics system which can treat the real world as an interactive network composed of agents subject to the satisficing principle. We will now confirm the features of this system.

There is no contradiction between the microbehavior and the macrovariables. This system is constructed in a completely bottom-up way, and all macrovariables are the result of the summation of each agent's behavior over a certain period.

The system is compatible with the irreversible and autonomous behavior of each component. Each agent can adjust his or her output regardless of whether the system is in equilibrium or not, based on his or her local information as input and referring to his or her own satisficing standard.

The interactions of these agents develop through irreversible time. Each output of one agent can directly affect that of others in so far as his or her output forms one item of input for other agents, and can also affect others indirectly in so far as the outputs of the affected agents are items of input for further agents. This makes it possible to completely comprehend the interactions of economic agents in irreversible time.

Since it has the same formulation as a single-layered, mutually connected neural network, this system also has many locally stable equilibria. This fact ensures some robustness in the system. Of course, even at these equilibria an agent does not optimize his or her utility function. An equilibrium simply means the situation where no agent makes any adjustment behavior.

This system is a loosely coupled system in the sense that agents refer only to their local information, and the response from each adjusting output comes after a time lag of at least one period. Such a loosely coupled system is referred to as a robust real system. In the case of a rigidly coupled system, each change of component requires an immediate reaction from every other component, so the behavior of the system tends to be unstable. Therefor the system tends to be vulnerable against shocks from the outside, and sometimes it does not work. Zadeh (1973) discusses the difficulty of formulating a large and complicated real system by rigid equations. However, in the case of a loosely coupled system, the robustness of the system is increased by localized shocks to each component. At the same time, the possibility of local and autonomous behavior from each component makes it possible to respond quickly to a change in the environment (see Ashby 1960; Weick 1976; Tanaka 1981).

4.4 Buffers for the Survival of Components

We now need to explain why we put b_j in our formulation. To follow the satisficing principle means to operate a thresholding process to the weighted summation of the input information, and to connect this result to the adjustment of behavior as the output. Of course, this is not an optimizing adjustment but a routine adjustment, and the size of the adjustment is limited, so it often occurs that an input which will necessitate an adjustment cannot be processed appropriately in a small number of adjustments. It is also possible that a further input will mean further adjustment. Hence, the satisficing agent should have the facilities to be robust enough to provide a buffer against any deviation from the "normal" state for lengthy periods.

Considering the size and complexity of the economic system in relation to each agent, the information input to each agent easily becomes too large for his or her processing capacity. Therefore, in Sect. 4.1 we severely limited the referring area. However, this may increase the possibility that the input information contains "noise" or an inappropriate adjustment signal. Hence, the agent also needs a buffer to cope with such noise. We have no space here to discuss this buffer in its concrete form—typically money or stock—in our economy. On stock as a buffer and its relation to the behavior of the whole system, see Morioka (1998).

4.5 Good Handling as a Sociotechnology

Based on the architecture above, to get good handling of the policy applications we need to group agents who can be seen to show the same kinds of behavior patterns as a macrounit. This operation makes it possible to treat the behavior of the economic system as an interactive system composed of these macrounits, and to decrease the number of variables in our application model.

There are two ways to approach this operation.

1. Agents can be put into groups according to their input/output processes. For example, in Japan the Goto-barai are special days on which deliveries and payments are normally made (perhaps a monthly payment on the 5th, 10th, 15th, 20th, 25th, or 30th). If this timing is synchronized, we can form a macro input/output unit of all agents in the same group.

2. If each of these groups is regarded as an interactively connected network, we can use statistical mechanics to treat a macro unit as a function which returns the average output (which is in response to the average outside stimulation) to its components as the next input. The unit is thus provided with the connection strength parameter (see Yoshida 1997 Chap. 5).

Of course, even after such a process, there are still too many groups with different behavior in our economic society. Therefore, we sometimes have to limit our attention to part of the system and carry out a subinteractive-system analysis. The best-known and simplest example of such an analysis is multiplier process analysis, as described in Sect. 3. If the behavior of all other groups is taken as being constant, this analysis concentrates on the interaction between two macrounits, i.e., salary-earning households and the entrepreneurs who produce consumption goods, and investigates the effects of their policies. This approach is limited by the narrowness of its time-sight and the field it covers, because it is valid only before the reactions of other groups are known. To develop a policy application model without these limitations, we have to add other macrounits and make wider interactive-network models. In expanding our application model, it should be noted that the results will differ widely according to the relative reaction speed of the macrounits.

In neoclassical economic development models, the behavior of the sector producing capital goods (investment) is considered with an assumption that the reaction between the sector producing consumption goods and wage-earning households has already been completed, i.e., the IS balance condition. However, there is no reason why we should assume that any change in the sector producing capital goods occurs after the sector producing consumption goods reaches an equilibrium. Therefore, we have to take their relative reaction speed into account, and have to design an economic theory based on an interactive system which can show the variations in results appropriately according to their relative speeds. However, economics based on equilibria, and without such a design, is forced to consider the behavior of the slowest adjustment sector and assume that the other sectors have already reached their equilibria.

We now give our fundamental equation for a macroapplication model.

$$X_j(t+1) = A_j \cdot F_j\left(\sum W_{ji} X_i(t) - S_j + Z_j(t), B_j\right)$$

where $X_j(t)$ is the adjusted output of the j-th macrounit in period t, W_{ji} is the average weight needed to estimate the adjustment to the output of the i-th macrounit as the input for the j-th macrounit, S_j is the average satisficing standard of the j-th macrounit, $Z_j(t)$ is the average stimulation to the j-th agent from outside the system, A_j is the parameter of the average maximum possible adjustment, and $F_j(\cdot)$ is an increasing function whose minimum/maximum value is ± 1 and whose output is 0 when the absolute value of the input value does not exceed B_j. This B_j reflects the average buffer size in each macrounit.

4.6 An Illustration by Some Simulations

We now examine some simple cases using this model. First, we consider the simplest interaction between the macrounit of entepreneurs, producing consumption goods and the household macrounit. The former adjusts its average production and employment level depending on whether revenue exceeds costs (including a "normal" margin) or not, and the latter decides its consumption level according to its income level in a regular pattern. From any initial level of the production of consumption goods, the entrepreneurs macrounit naturally converges to the same global result (Fig. 2a). This is the case of an effective demand theory.

We now add more macrounits, i.e., a macrounit of entrepreneurs producing capital goods and a macrounit of investment entrepreneurs. The former adjusts its production level according to its extra profit from the macrounit producing consumption goods, and the latter adjusts its investment according to the movement of gross national product (GNP). As the natural result of their reactions, the system follows different paths according to the initial production level of the unit producing consumption goods and reaches different equibria (Fig. 2b). Furthermore, the system increases the diversity of its paths and destinations if we further add an asset-trading macrounit (Fig. 2c). For each case, if we introduce a buffer in the adjustment reaction of the macrounit producing consumption goods, the variation resulting from the

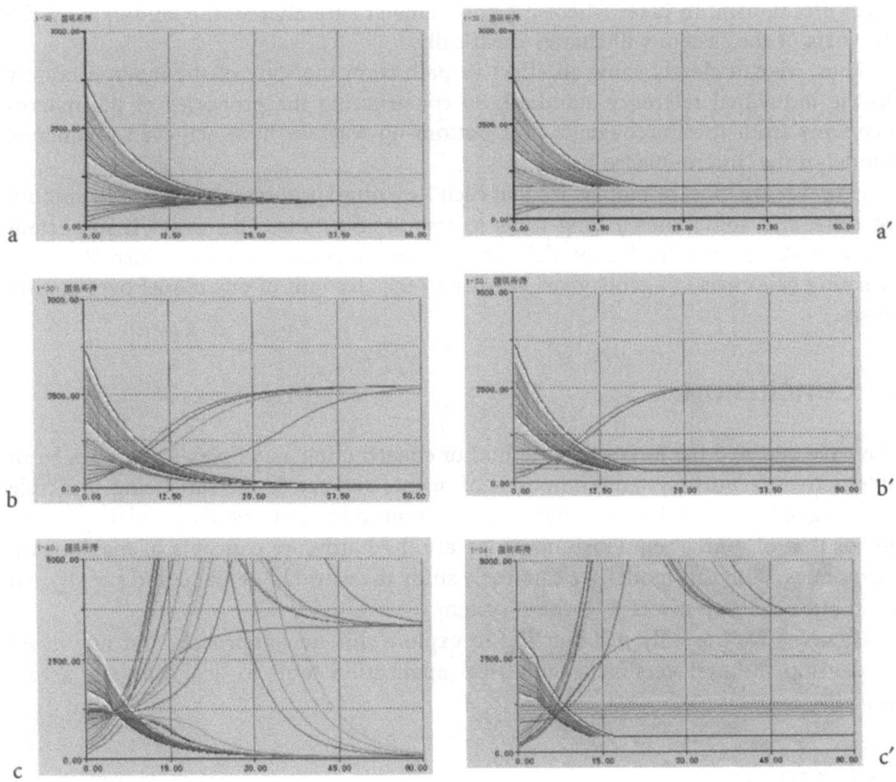

FIG. 2. The interaction among macrounits with or without a buffer

starting data decreases, and in this way the robustness of the system is increased (Fig. 2a′,b′,c′).

4.7 How Can We Express the "Micro–Macro Loop"?

Some authors have referred to the "micro–macro loop" (Shiozawa 1997), which means the interaction between the activities of each agent and the behavior of the whole economy. We can easily show the "micro–macro loop" model here based on the model given above and expanded by the multilayered feedback mechanism reported by Mesarovic et al. (1970).

The "satisficing" standard or the "normal" concept of each agent is changed if it has not been realized for a long time. In formalizing the satisficing agent, we need some further parameters. These parameters show routine adjustments, and were fixed in the formulation given above for simplicity, but some of them should be expected to change over comparatively short periods, and others should be expected to change only in the long term. We can easily add such a learning mechanism to our agent model by assuming a lower-layer feedback loop for the frequently altered parameters, and a higher-layer feedback loop for the basic parameters, in the multilayered feedback mechanism. In cases where many parameters are altered, the landscape of the equibria of the system will change drastically.

Thus, we can clearly show an effective path from the state of the macroeconomy to the individual reference standard. By constructing the processes of the macroeconomy from the microagents in a bottom-up way, we have achieved a concrete model of the "micro–macro loop."

However, we must re-emphasize that each "learning" process takes a certain amount of time, and sometimes a long time, in irreversible time. This point is important when making an effective model for the real working world, because the slow learning processes of agents gives validity to the strength of our model over several years.

5 Conclusion

We have outlined the Keynesian method of constructing a complex economics as an interactive network system composed of agents subject to the satisficing principle working in historical time. Our model does not need the concept of optimizing behavior or that of market equilibrium, which are the hard-core concepts of mainstream economics. With this model, we can freely study the world of demand and supply, and treat our economy as a real working system.

However, we are only just starting to explore this new approach, and now need to develop the first version of empirical application tools to deal with the actual world.

References

Ashby WR (1960) Design for a brain, 2nd edn. Chapman & Hall, London
Eichner A (1979) A guide to post-Keynesian economics. Macmillan

Hirai T (1987) A study of Keynes: from the *Treatise on Money* to the general theory. Tokyo University Press, Tokyo
Keynes JM (1971) Collected writings of John Maynard Keynes, vol V. A treatise on money. I. The pure theory of money (1930). Macmillan, London
Kregel JA (1975) The reconstruction of political economy: an introduction to post-Keynesian economics, 2nd edn. Macmillan
Mesarovic M, Macko D, Takahara Y (1970) Theory of hierarchical multilevel systems. Academic Press, New York
Morioka M (1998) Input–output structure, forcasting of the sales, and buffering stock. Econ Rev 161(1):108–132
Negishi T (1980) The micro-theory of the Keynesian economics. Nihon-Keizai-Shinbunsha, Tokyo
Robinson J (1971) Economic heresies: some old-fashoned questions in economic theory. Basic Books, New York
Shiozawa Y (1990) A study on market order: from anti-equilibrium to complex system. Chikuma-Shobo, Tokyo
Shiozawa Y (1997) The result of complexity. NTT, Tokyo
Tanaka (1981) The theory of loosely coupled system. Organ Sci 15(2):59–75
Weick KE (1976) Educational organization as loosely coupled systems. Admin Sci Q vol 21, March
Yoshida M (1997) Keynes—from the “historical time” to the economics of complexity. Nihon-keizai-hyouronsha, Tokyo
Yoshida M (1988) Theoretical structure of Keynes’ *Treatise on Money*. Econ Rev 142(5/6): 91–108
Zadeh LA (1973) Outline of a new approach to the analysis of complex systems and decision process. IEEE Trans Syst Man Cybern 3:28–73

Why Should Economists, Including (Old) Institutionalists, be Interested in Critical Realism?

Tony Lawson

Summary. Modern economics has many problems. The angument is made that these can best be overcome by way of an ontological turn, i.e., by reorientating the discipline in the direction of realist social theorising.

Key words. Ontology, Social theory, Mainstream economics, Institutionalism

1 Introduction

The primary objective of this chapter is to argue that the discipline of modern economics has much to gain by reorienting itself in the direction of realist social theorizing. In particular, I wish to suggest that the discipline can profit significantly by taking on board some of the results of a specific project in realist social theory systematized as critical realism. In making my case, I will inevitably be highlighting certain failings of alternative approaches against which realist social theorizing is contrasted. Most of the criticisms will concern the mainstream project of the modern discipline, but certain fundamental ones will be seen to apply also to aspects, or strands, of heterodox projects such as that systematized as old (or Veblenian) institutionalism.[1]

2 Defining my Terms and Explaining my Orientation

Let me start by defining some of my basic categories and explaining my orientation. By social theorizing, I understand all systematic and sustained attempts to illuminate social reality. By social reality, I understand that domain of phenomena whose existence depends at least in part on human (intentional) agency. By realist social theorizing, I understand social theorizing in which ontology, i.e., the study of the nature and structure of social reality or being, is explicit and sustained. Critical realism in economics, then, is a specific project of this latter sort.

Faculty of Economics and Politics, University of Cambridge, Sidgwick Avenue, Cambridge CB3 9DD, UK

[1] I engage somewhat with this project or tradition here, of course, not because I am especially critical of it (on the contrary), but because I have been invited so to do, old institutionalism being the tradition which underpins the conference for which this work was originally invited and produced.

Why the concern with ontology? As I say, the attention given to ontology, to (the study of) the nature of being or existence, is a key feature of critical realism. Indeed, the emergence of critical realism in economics has been referred to by some as an ontological turn in modern economics. But what is the point of it?

Imagine a situation in which a pneumatic drill has been found successfully to make a hole in the road, and on the basis of this success is selected for a second task: that of cleaning a glass window. Clearly we know this to be a preposterous decision, and are not at all surprised when the drill causes the glass of the window to break. But why do we suppose that the decision so to employ the drill can be recognized as mistaken, even in advance of the outcome? Precisely because we know something of the natures of both the glass and the proposed method of illumination (the drill) to know that the one is not well suited to the other; given what most of us know about the nature of glass, we would expect something softer, like a cloth, to do a better job. Now just as we can determine the natures of specific objects such as glass windows, so, I believe, we can obtain insights into the nature of the basic subject matter of a field of study. To do so is to engage in ontology. Thus, in an analogous manner to matching the cleaning tools to the window, so, I believe, by way of ontological reasoning, we can assess the relevance of proposed research procedures to illuminating social reality.

This rather stark analogy hopefully illustrates something of why I take ontology to be essential to social theory, at least in principle. As it happens, it is the case that modern economics is in a state of some disarray whilst remaining almost wholly neglectful of explicit ontological analysis. My thesis, defended below, is that it is ultimately this ontological neglect that accounts for the discipline's poor showing. This, in my view, is why economics in particular warrants an ontological turn, and why, especially now, it can benefit significantly from a reorientation in the direction of realist social theorizing.

The observation that the modern discipline of economics, or at least the hugely dominant mainstream project, is not in a healthy state (it performs poorly on its own predictive and explanatory criteria, is riddled with theory/practice in-con-sistencies (Lawson 1997a), and has no clear rationale or direction) is, I think, widely made, even by reflective members of the mainstream themselves,[2] and is no longer especially contentious (see, for example, Bell and Kristol 1981; Kirman

[2] To consider just one assessment by a prominent "theorist," Rubinstein in his tribute to Nash's contribution to game theory, expressed puzzlement as to the mainstream "economic theory" project's ultimate purpose and direction:

> The issue of interpreting economic theory is, in my opinion, the most serious problem now facing economic theorists. The feeling among many of us can be summarized as follows. Economic theory should deal with the real world. It is not a branch of abstract mathematics even though it utilises abstract tools. Since it is about the real world, people expect the theory to prove useful in achieving practical goals. But economic theory has not delivered the goods. Predictions from economic theory are not nearly as accurate as those by the natural sciences, and the link between economic theory and practical problems, such as how to bargain, is tenuous at best. Although I have never heard an economist seriously claim that the Nash bargaining solution is a good predictor of bargaining in real markets, this solution is a standard tool in modelling interactions among negotiators. Economic theory lacks a consensus as to its purpose and interpretation. Again and again, we find ourselves asking the question "where does it lead?" (Rubinstein 1995 p. 12)

1989; Lawson 1997a; Leamer 1978, 1983; Leontief 1982; Rubinstein 1995; Wiles and Routh 1984).

My assessment of the cause of this disarray, as I say, is ultimately that of ontological neglect. I aim to demonstrate that certain dominant methods are adopted not because reflection upon the nature of social material has indicated their potential relevance to illuminating the social realm, but because (like the drill) they are perceived to have been useful in some really quite different domain. Specifically, the majority of methods used by modern-day economists are procedures of formalistic modeling, perceived to have been successful in certain natural scientific contexts. I want to argue that the latter perception, even if correct (and I shall be suggesting that it is not), would, by itself, not be enough; there is no automatic guarantee that tools found to be appropriate for dealing with one sort of material must also be appropriate for another that is potentially quite different. The relevance to analyzing the social realm of tools developed in a natural scientific context, however, can be explored by way of ontological analysis, and by way of such analysis, I shall argue, the legitimacy of the use of formalistic methods can be shown to be limited at best, and possibly not of relevance at all.

I might also add that ontological analysis also bears implications for the relevance of strands of institutionalist and/or evolutionary analysis (if amongst aspects of other heterodox projects), a matter to which I shall turn explicitly in due course.

3 Philosophical Argument and Ontology

How, though, do we do ontology? And how does it compare with other prominent forms of philosophical analysis? Much recent philosophy of science concerns itself with attempts to justify scientific practices already assumed at the outset to be rational; it attempts to substantiate an unquestioned a priori premise of their rationality. This form of philosophy not only tends to be circular,[3] and, when applied to economics, ignores the fact that the discipline is in a state of disarray, but also encourages the implication that, because it makes no difference to actual practices, all philosophy of science or methodology can be safely ignored by the practicing scientist. That this latter situation characterises much of economics has already been observed by Blaug:

> Too many writers on economic methodology have seen their role as simply rationalizing the traditional modes of argument of economists, and perhaps this is why the average modern economist has little use for methodological inquiries. To be perfectly frank, economic methodology has little place in the training of modern economists. (Blaug 1980 p. xiii)

Of course, it is nothing new for economists to start from the assumption that forms of behavior are optimal in some way, and to work backwards to specifications of conditions against which the assumption of optimality allows already observed behavior to be deduced. Such circularity is endemic to modern mainstream economic exercises. As such, it is not surprizing that economists should apply the same logic to any philosophy-of-science endeavor. Of course, ex posteriori lay agents do appear to be

[3] Philosophical argument of this sort tends to uncover conditions which, if they held, would render actual practices rational, and then to assert for that reason alone that such conditions must hold.

successful on a day-to-day basis in negotiating social reality. However, whilst this observation does not yet licence the assumption that most people continually optimize, it remains the case that in the realm of academic economic research, there are as yet few signs of illuminatory successes at all. Here, of all spheres, an a priori assumption of optimality is question-begging, and Blaug is correct to be critical of the conventional scenario.

Blaug's own alternative starting point has been to emphasize (Popperian) injunctions formulated largely through reflecting on the conditions of natural science. However, these formulations have repeatedly been rejected by practising economists as not really practical in the social realm (see, for example, Leamer 1978, 1983). In the light of this realization, various metamethodologists, in particular, have called for the abandonment of methodology more or less altogether (see, e.g., McCloskey 1986; Weintraub 1989), thereby reinforcing the antimethodology sentiment already prevalent.

This, then, has been the context in which critical realism has emerged. The discipline of economics is in something of a bad way, and philosophical argument continually receives something of a bad press. How is possible to do any better? How, in particular, can ontological analysis proceed.

4 The Method of Critical Realism

The starting point for those contributing to the project of critical realism has been not that scientific practices are necessarily rational (or irrational), but that they, and indeed all practices, are *intelligible*. That is, the contention sustained is that, whether actual practices are successful or not on their own terms, they have explanations; there are conditions which render the practices actually carried out (and their results) *possible*. Thus, a central strategy has just been to seek to explain (aspects of) certain human actions, and to identify their conditions. This, of course, is precisely an exercise in ontology.

5 Transcendental Reasoning

To start from the intelligibility of a form of life should not be contentious. After all, we do all assume that at least some aspects of the world, including the practices of others, are intelligible. It is difficult, for example, to imagine anyone bothering to attempt to read and understand these lines who supposes otherwise.

The form of reasoning that takes us from generalized practices (or aspects of them)—or more generally from widespread features of experience—to their grounds or conditions of possibility is the *transcendental argument*.[4] The transcendental argument or transcendental deduction is thus clearly a special case of the *retroductive argument*, where the latter moves from conceptions of specific phenomena at any one

[4] This accounts for the use of the transcendental realist label for specific insights achieved by Bhaskar (1978) in proceeding in this way in considering the natural realm. Of course, Kant used the same form of argument, but to ask questions about us, and the presuppositions of our knowledge of being (hence his transcendental idealism), but there is no need to restrict transcendental argument to the individualist and idealist mode employed by Kant.

level to hypotheses about their underlying conditions or causes (see, e.g., Lawson 1997a, especially pp. 24, 212–15).

Any results achieved by way of transcendental reasoning are clearly conditional—upon the human practices selected as premises and our conceptions of them, as well as upon the adequacy of the transcendental argument employed.

Moreover, it is clear that philosophy so conceived, i.e., as method turning centrally upon the transcendental argument, considers the *same world* as the sciences, and indeed serves to complement their results. However, it proceeds on the basis of pure reason (albeit exercising it always on the basis of prior conceptions of historically rooted practices), and produces (fallible) knowledge of the necessary conditions of the production of knowledge.

6 Modern Economics

Let me consider the situation in modern economics from the point of view of ontology. I think there is little doubt that the dominant method is that of formalistic modeling. This comprises most of microeconomics, macroeconomics, and econometrics. Indeed, formalistic modeling is the essence of the mainstream economics project in modern times, and is even portrayed by that project as the core of the whole modern discipline.

Now all such formalistic modeling endeavors warrant, as a condition of their successful widespread application, that closed systems—i.e., systems in which regularities (deterministic or probabilistic) of the form "whenever event x then event y" occurs—are ubiquitous. Forms of explanation that require the formulation of event regularities of this sort are called deductivist.

If it had turned out that formalistic deductivist modeling practises had proven widely successful in economics, we could pose the ontological question with respect to those practises; we could transcendentally deduce the conditions for their successes. One clear condition of the possibility of such successes—*had they occurred*—is precisely that in the social realm sufficiently strict event regularities had been found to be ubiquitous.

However, so far, formalistic modeling methods have not been found to be very successful. This precisely reflects the fact that sufficiently strict or stable event regularities are yet to be uncovered. Of course, mainstream modelers mostly regard the lack of success with this direction of research as a temporary phenomenon. Typically, the main quasimethodological conclusion to be drawn is merely that it is necessary for modelers to search somewhat harder.

In their seemingly undaunted perseverance with this direction of research, the protagonists of the mainstream approach are motivated in their thinking by (i) a desire to be scientific in the sense of natural science, coupled with (ii) a conviction that natural science, indeed all science, takes the form of formalistic deductivist modeling. Stuck with such desires and beliefs, mainstream practitioners are convinced that there really is no alternative way of proceeding.

Thus, for example, we find Hahn criticizing dissenters from the mainstream approach for not realizing that the deductivism he recommends is indeed the universal approach:

> Opponents of [economic] theory often argue that it is tautologous because it consists of logical deductions from axioms and assumptions. If one is kind to such critics one interprets them as signalling that they do not care for these axioms or these assumptions. In any case all theory in all subjects proceeds in this manner. (Hahn 1985 p. 9)

In addition, Allais supposes that the existence of regularities is the phenomenal situation in economics, a state of affairs which explains, as he sees it, why economics is a science resting on the same general principles and methods as natural sciences such as physics:

> The essential condition of any science is the existence of regularities which can be analyzed and forecast. This is the case in celestial mechanics. But it is also true of many economic phenomena. Indeed, their thorough analysis displays the existence of regularities which are just as striking as those found in the physical sciences. This is why economics is a science, and why this science rests on the same general principles and methods of physics. (Allais 1992 p. 25)

Now in the absence of ontological analysis, it may seem, prima facie, not unreasonable continually to expect that success is just around the corner for formalistic economics; it is conceivable that an enduring faith (for that is all it is) in the relevance of deductivist explanation, and specifically the reliance upon formalistic modeling methods, will be vindicated with time. However, ontological analysis gives reason to suspect otherwise in providing what is, I believe, a more compelling alternative explanation of the mainstream formalistic project's poor showing so far. Moreover, it indicates that alternative ways of proceeding are likely to prove more profitable. Let me elaborate this set of claims.

7 Some Transcendental Deductions

I start by questioning that already noted presumption of so many economists (captured by the statements of Allais and Hahn above) that formalistic deductivist modeling procedures, and so the elaboration of event regularities, are central, indeed essential, features of natural science.[5] As I say, even if this assessment were correct, by itself it would not in logic justify the observed (uncritical) perseverance with deductivist methods in the social realm, no matter how strong the desire to be scientific in the manner of natural science. However, if this presumption about the nature of natural science can itself be shown to be false (and we will see that it can), the usual scientistic motivation for persisting so uncritically with (so far unsuccessful) formalistic methods in economics will be seen to be, even on its own terms, quite misguided.

---

[5] I might also note that the supposition that closed systems (i.e., those supporting regularities of the form "whenever event x then event y") are essential to natural science also appears to be accepted by a high proportion of the few economists who have recognised their limited relevance to the social realm, economists who have concluded thereby that economics cannot be a science (see, for example, Hayek 1942–44; Hicks 1986).

8 The Experimental Natural Sciences

Here it is useful to consider the intelligibility of aspects of those practices bound up with such event regularities as do occur in the natural realm. Specifically, I want to make two observations. The first is that event regularities which are considered significant in science are, in fact, rather rare and largely associated with a specific form of human scientific practice. That is, outside astronomy, most of the event regularities considered to be of interest to science are produced in conditions of experimental control. Second, the results obtained in experimental work are regularly successfully applied in conditions in which event regularities do not occur.

How do we render these aspects of human practices intelligible? What are their ontological conditions? Note first that these observations are already problematic for the deductivist conception, for which the elaboration of event regularities is regarded as essential. The recognition that actual event regularities are in fact highly restricted in their occurrence severely undermines the idea that science is universally applicable; the realization that such regularities are largely confined to conditions of experimental control bears the implication that any "laws of nature" depend on us (for human intervention is required in setting up the experimental conditions) and, most significantly of all, the fact that experimental results are successfully applied in conditions where event regularities do not occur is found to be quite inexplicable.

The noted observations concerning experimental practices can be rendered intelligible, but in order to do so it is necessary to recognise explicitly a deeper level of reality to the sorts of event and states of affairs connected in "whenever this then that" deductivist formulations: there is more to reality than the items which comprise possible objects of experience. Rather, we need to recognise that behind the course of actuality, behind actual events and states of affairs, lies a realm of structures, causal powers, mechanisms, tendencies, and the like; conditions which produce such events and states of affairs as can be experienced. Let me describe this ontology, and explain what I understand by categories such as power and tendency, before indicating how the conception defended can render the noted observations intelligible.

Consider an aspirin. It (usually) appears white and either solid or powdery. However, it also contains an unobservable chemical structure. Because of its intrinsic chemical structure, it has the power to relieve a headache (in most people). However, the power possessed by a given aspirin, although real, may never be exercised; the aspirin may always stay inside its bottle. Even if taken, the aspirin may work as a mechanism to relieve the headache, and yet the headache may actually get worse. This may happen, for example, if the person with the headache not only takes the aspirin, but simultaneously drinks a bottle of *Sake*. The idea of a mechanism having an effect which may be different to the actual outcome is captured by the category of a tendency. It is the idea of *continuing activity* (as distinct from enduring power) that the notion of a tendency is designed to express. It refers not to a counterfactual scenario, to something that would happen in different circumstances, but to a *transfactual* mechanism, one that is really in play and having an effect whatever the actual outcome. The taking of the aspirin sets in train a real tendency to relieve the headache, but the final outcome is dominated by the countervailing tendency for the pain to get worse as a result of the consumption of a bottle of Sake. Here, the aspirin is having its effect whatever the outcome. The gravitational tendency is acting on the pen in my hand, and will continue to do so whether I continue writing with the pen, throw it in the air, or drop it into a vacuum.

I suggest that it is such an ontology—of structures, powers, mechanisms, and tendencies—that is required to make sense of the observations noted above. From this perspective, the limited occurrence of event regularities is explained by an open structured world of tendencies and countervailing tendencies, into which scientists intervene with their controlled experiments designed to isolate individual stable mechanisms or tendencies from others that may have countervailing effects.

First we can see that the restriction of event regularities to the experimental situation no longer places a limit to the practice of science. For science is now properly recognized as the process of seeking out not correlations in surface phenomena such as events, but the causes of surface phenomena, and the experiment is seen as merely one set of procedures that can help in this process. Second, the experiment is not interpreted as producing laws of nature (understood by mainstream economists as, or as necessitating, event regularities), but merely as producing those conditions under which causal mechanisms or tendencies (now recognized as the proper objects of any laws of nature) are empirically identified. There is nothing counterintuitive about this. Third, the fact that the results of the experiment are successfully applied in conditions where event regularities do not hold is quite intelligible, for the results of experiments do not concern event regularities per se, but transfactually acting mechanisms that are experimentally empirically identified. Thus, the gravitational tendency may be empirically identified in the controlled experiment, but it acts both inside and outside the experiment alike, even on the autumn leaf as it flies over roof tops and chimneys.

Already, then, our ontological reasoning bears implications for economics. Any desire to be scientific in the sense of natural science does not after all necessitate the relentless pursuit of formalistic deductivist methods in the face of repeated failure. Rather the question of naturalism—whether economics can be scientific in the sense of natural science—must now be posed somewhat differently. Specifically, it takes the form of questioning whether social phenomena at any one level can be explained in terms of their causes lying at another. The answer to this question, of course, can be determined only by looking to the social realm directly.

9 Ontology and Social Reality

I start my development of a social ontology with two observations concerning human practises. The first is that econometricians have yet to turn up many, if any, event regularities of the degree of strictness for which they are looking. Second, human beings, all the same, do seem to be rather successful at negotiating social reality. How do we make sense of these generalized features of our experience?

One explanation of the first observation—seemingly rejected by most practising modern economists—is that social regularities that are both of the sort being sought and of interest, just do not occur. This is an explanation that, I believe, will prove compelling in due course. The second observation presupposes, I think, that the social world is predictable none the less. Let me elaborate, and examine the conditions of possibility of such a situation.

First, let me recall that by the social realm I mean that domain of phenomena whose existence depends at least in part on us. Now it is a generalized feature of experience that children take a while to learn their way around a particular social system, as often do visitors coming from an alternative system that is culturally quite different. Yet most adults acting within their own locale seem more than competent in negotiating their way around. Such observations suggest both that social reality is everywhere complex and yet, in many of its aspects at least, is also everywhere something that can be, and is, well known. The question, then, is not *whether* successful social theorizing is possible, but *how* in fact it is done.

As I say, it is unimaginable that the multitude of continuous human interactions could be successfully performed if the social realm was not at least predictable much of the time in some of its aspects. A precondition of this is the existence of social rules, procedures of action laying down how we ought legitimately to act in various situations. This could be (indeed is) a transcendental inference, but in truth we already know the inferred conditions to be so; we already know that social life is rule-conditioned. Social rules are best conceived of as *generalized procedures of action*, procedures that, under a suitable transformation at least, can be expressed as injunctions of the form: "if *x* do *y* under conditions *z*." For example, "if you wish to speak at a crowded seminar, hold your hand up when in twentieth century Britain." The stipulation "under conditions *z*" will often be dropped in any explicit formulation, but will always be implicated albeit unacknowledged. All action, for example, takes place over limited regions of time and space.

If we already know that social reality is in part constituted by social rules, there are various aspects of them and their own conditions that warrant elaboration. In particular, social rules should be recognized as being ontologically distinct from social practices. This follows once we note, and enquire into (transcendentally deduce) the conditions of possibility of, the fact that practises governed by rules are not always, or on average, in conformity with them. The (intentional) act of rebelling requires as much knowledge of the rules as does that of conforming. Currently, motorway drivers in the UK *mostly* drive at a speed above that laid down in law, albeit usually not faster than others for fear of getting caught out, and workers contemplating industrial action frequently threaten to work to rule. Making sense of the fact that rules and the practices they condition are often (or even sometimes) so much, and possibly systematically, out of phase requires that we recognize the two aspects, rules and the practices they condition, as connected, but ontologically distinct, types of thing. The ontological distinction between social rules and practises is a (transcendentally inferred) necessary condition of the possibility of the latter influencing, whilst often being out of phase with, the former.

Note that this ontological distinction, with rules comprising a feature of social structure facilitating, but irreducible to, human practices, is already enough to affirm the possibility of naturalism. That is, we can already accept that the study of social phenomena can be scientific in the sense of natural science; that social analysis can move from phenomena at one level to their causes at a different "deeper" level.

However, more can be established in this manner. A further generalized feature of experience is that (clearly rule-governed) practises are segmented: women, for example, often, and systematically, follow different practises (and so rules) to men, teachers to students, employers to employees, landlords to tenants, and so forth.

Further, any given set of *continuing* practises in a specific context can be performed by different individuals. For example, each year I give lectures, set essays and exams, and so forth, and students are obliged to listen to the lectures, and produce essays, and take exams, but the set of specific individuals enacting these student tasks changes from year to year.

Making sense of these observations requires that we recognize, as a condition of their possibility, that the social world consists in part of differentiated social positions. It is positions—into which individuals essentially slot—such as those of student or teacher or employee that endure and have rules, and so have tasks and obligations and prerogatives attached to them, and it is because there are a multitude of different positions, that different practises are followed by the members of any societal group.

However, the practices followed by individuals in different positions are frequently more than just different; very often they are *other-oriented* in very systematic ways. The practices of students and teachers are oriented toward each other, as are the practices of employers and employees, or of parents and children, or of landlord/ladies and tenants, etc. Making sense of this further observation requires that we recognize, as its necessary condition, that the social world is also in part constituted by social *internal* relations. Internal relations are those whereby the *relata* are what they are, and/or are able to do what they do, in virtue of the relation in which they stand. Internal relations hold, for example, between a magnet and its field, or between the teacher and the student; in each case you could not have the one without the other. Note that in the social realm, it is the internal relations between the positions that are of primary significance.

In short, we are able to make sense of various generalized features of certain human practises by transcendentally deducing their conditions, and in so doing we are led to a definite conception of social reality. According to this, social reality is structured (irreducible to actualities such as social practises and other events) and segmented (or class divided), and consists in social rules, relations, and positions, amongst other things.

10 Some Further Ontological Elaborations

Notice too, that various further conceptions can be built out of these (transcendentally deduced) categories. All social systems and collectivities can be recognized as ensembles of networked, internally related, positions with associated rules and practises. This applies to the state, schools, hospitals, trade unions, the household, and so forth. Subdistinctions can be made. A social system can be interpreted as a structured process of interaction, an institution as a particular social system/structure that is relatively enduring and regarded as such, and a collectivity as an interrelated set of social positions along with their occupants, and so forth (see Lawson 1997a pp. 165–6).

These categories also provide the basis for various other conceptions, including a theory of situated rationality (Lawson 1997b). The interests of any given individual at any moment, as well as their possibilities for action, will depend to an extent upon the internally related positions in which they are located. Of course, we all always stand in a large number of positions (as parents, children, immigrants, indigenous, old, young, teachers, etc., etc.). Hence there exist possibilities of conflicting as well as unrecognized, individual as well as collective or shared, interests, and so on.

More can be inferred. Because social structure (in virtue of being social) depends on us (i.e., upon always potentially transformative human agency), structural deterministic accounts have to be rejected. Because ex posteriori much social structure is found to preexist any set of us and make a difference—we could, for example, not speak as we do without the prior existence of language, drive safely on motorways without knowing the already existing highway code, etc.—we can infer that social structure is both relatively autonomous (it preexists our current acts) and real (it makes a difference to what is possible). The fact of social structure making a difference to human action, and therefore its ability to do so, i.e., its possession of irreducible causal powers, presupposes an account of emergence, just as its dependence on always inherently transformative human agency establishes its inherently dynamic mode of being.

Thus we arrive at a conception of social reality as an emergent realm dependent upon, though irreducible to, inherently transformative human agency, and consisting of stuff that is intrinsically dynamic and highly internally related, amongst other things.

11 Some Implications for Analysis

Ontological analysis of this sort can and does carry numerous methodological implications. These bear both on potentially fruitful possibilities for research practice, and on limits to what can be achieved. I have emphasized the former on numerous occasions[6] (see, for example, Lawson 1997a Chap. 15; 1999). Here I want to take a somewhat more critical orientation, focusing in particular on implications concerning suspect practises prevalent in the mainstream project, but seemingly evident in certain strands of heterodox traditions like (old) institutionalism as well.

11.1 Accounting for the Modern Mainstream

As a first consequence of the above ontological excursions, I am now in a position to explain the situation of modern economics. When the conditions of the experimental natural sciences were examined above, I observed that event regularities are produced where intrinsically stable mechanisms are insulated, and thereby empirically identified. The event regularities are basically relations between triggering conditions and the effects of (insulated) mechanisms. The conditions of experimental successes of this sort are thus found to be (i) that reality is open and structured (allowing scientists to intervene to close the system by isolating particular mechanisms), and (ii) that at least some structures and mechanisms are indeed isolatable (reality is so differentiated) and also intrinsically stable (especially if the same event regularities are to be produced in repeated trials).

[6] Suffice it to say that with the social realm not comprising isolatable manipulable atomistic bits, social scientific explanation is necessary backward looking, and given that it is not possible directly to control any causal factors, including background ones, identification of specific causal factors requires selecting a relatively homogeneous environment. Such considerations, I have elsewhere argued, necessitate a form of *contrastive explanation*.

Now it is no doubt an implicit recognition that conditions such as these must somehow be satisfied (if event regularities are to be elaborated) that encourages mainstream economists to respond to their modeling failures either by including ever more variables in the formulations (attempting to isolate a complete system), and/or by digging ever deeper for invariances (to uncover atomistic elements treatable as the sorts of intrinsically stable mechanisms isolated in experimental systems).[7] Ontological analysis, however, has now revealed social reality to be not only structured (allowing the possibility of a science of social phenomena), but also highly internally related, intrinsically dynamic, and dependent upon transformative human agency. As such, it is highly unlikely that there are atomistic (or otherwise separable intrinsically stable) bits of society that can be isolated (experimentally or otherwise) to produce event regularities. On the contrary, the fact of the social domain's highly internally related nature seems to suggest that, while economics can yet be a science in the sense of natural science, the mainstream tradition, with its a priori insistence on formalistic deductivist modeling, has unthinkingly generalized that one aspect of natural scientific practise that possibly has hardly any relevance to the social realm at all. No wonder, then, that the modern discipline is in a state of disarray. It can with reason be said that this is due, in the main, to its continuing practice of ontological neglect.

11.2 Modern Institutionalist Thinking

Are heterodox traditions open to a similar form of criticism? Much less so, but even here there are some grounds for worry. At the conference for which the first draft of this article was originally prepared, a conference sponsored by those taking a Veblenian (old) institutionalist or evolutionary approach,[8] a large number of contributions were concerned specifically with advanced methods of nonlinear modeling. The only justification for this emphasis that I was able to discern was that the approach is considered to be modern or advanced, being facilitated in a meaningful way only recently through advances in computer technology.

Now it is true that arguing for an approach on the grounds merely of its being modern has an important precedent in the institutionalist tradition. In essence, indeed, at least at significant points, this is precisely the method of Veblen. Specifically, in asking why economics, at the turn of the twentieth century, was not yet an evolutionary science, and suggesting (erroneously) that inevitably it was soon to become one, Veblen nowhere argues the superiority of the evolutionary or "cumulative change" approach over any other. At least, he does not do so on ontological or metaphysical grounds.[9] In fact, Veblen takes the category of *metaphysical* to mean something beyond analysis, so that ontological or metaphysical preconceptions are

[7] For a detailed analysis of such responses, see Lawson, 1997a, Chaps. 7 and 8.

[8] A conference organized by the Japanese Association for Evolutionary Economics.

[9] I should add that I think Veblen is one of the foremost contributors to economics, and I hope it will not be considered uncharitable to criticize his writing here. Of course, no contribution is infallible. My point here is merely to identify possible weaknesses, and to caution against modern-day economists abducting these aspects in an overly uncritical fashion. The point is continually and critically to transform the better economic contributions, not to accept all aspects without question.

necessarily without evidential (or nonpragmatic) grounds. Thus, conceptions such as evolution and consecutive change are asserted to be without justification; "[the postulate of consecutive change] is an unproven and unprovable postulate—that is to say it is a metaphysical preconception . . ." (Veblen 1908 p. 33).

Rather, to the extent that the rise of the evolutionary perspective is supported by Veblen at all, it is so solely precisely on grounds of its *being modern*. This particular pragmatic criterion is evident in statements such as the following:

> The men of the sciences that are proud to own themselves "modern" find fault with the economists for being still content to occupy themselves with repairing a structure and doctrines and maxims resting on natural rights, utilitarianism, and administrative expediency. This aspersion is not altogether merited, but is near enough to the mark to carry a sting. These modern sciences are evolutionary sciences, and their adepts contemplate that characteristic of their work with some complacency. Economics is not an evolutionary science—by the confession of its spokesmen; and the economists turn their eyes with something of envy and some sense of baffled emulation to these rivals that make broad their phylacteries with the legend, "Up to date." (Veblen 1898 p. 57)

Of course, the nature of the (broadly Darwinian) evolutionary perspective in question is that it identifies groups/species/preconceptions that turn out to be relatively enduring, that "survive" or persist because they are found to be in some way appropriate to certain relevant aspects of their immediate environment. In the social, realm persisting or relatively enduring items, features upon which an evolutionary perspective encourages a focus, clearly include (by their definition) habits, and institutions, and the like. Hence, we find a considerable emphasis on such categories in Veblen's writing.

My worry, of course, is that this way of going about things, i.e., pronouncing in favor of an approach just because it is modern, encourages (as with all a priori approaches) a tendency toward ungrounded universalizing in theory development. In contrast, where a realist perspective as understood here is accepted, the relevant question for modern social theorists concerning the issues here being discussed is: given what we know about the nature of social material (derived from ontological analysis), what, if any, is the scope of legitimacy/relevance of any proposed approach. This remains the relevant issue: whether the methods in question share with those of Veblen either (i) the characteristic of being modern, or more specifically (ii) the characteristic of being primarily concerned with institutions and evolution.

To focus further upon (i), i.e., upon lending support for an approach just (or mainly) because it is perceived as modern, I should, of course, acknowledge that at any moment there will usually be competing interpretations of what counts as modern in science (although this does not affect the force of the argument). Thus, I do not mean merely to criticise present-day nonlinear approaches to modeling under this head. An alternative example of cutting-edge science currently on offer is provided by Mirowski, who is frequently interpreted as an institutionalist. Asserting that the "tyro and the average economist" are always concerned to know where the discipline is headed, and wanting to speculate upon an answer, Mirowski (1998), in a paper subtitled "economic agents as cyborgs," plumps for the cyborg science project, hoping and expecting economics to emulate those branches of biology, computer science, and operations research that are already perceived to be of this sort. Cyborg sciences are those dealing with hybrid forms that are usually part inanimate, part something else. Why does Mirowski recommend such a reorientation. The answer, it seems, is just because modern economics should avoid lagging behind developments in the physical sciences. He writes:

> I fully expect that economists will become more infatuated with the cyborg sciences in the immediate future; indeed, I think it inevitable. The reasons for this belief should now be a little clearer. Economics has persistently lagged behind the truly exciting developments in the physical sciences, only to take them up later in a distorted and distended version. (1998 p. 36)

It is true that Mirowski attempts to add support to this desired "recalibration of economics" (p. 16) by suggesting that "Every major transformation of the postwar orthodoxy in economics can be understood as a more or less direct function of a prior development in a related cyborg science" (p. 16). Although I believe this claim to be erroneous (see Ingrao and Israel 1990), this is not my point here. The postwar orthodoxy, just like Mirowski, is neglectful of ontology. Mirowski's arguments, like many postmodern contributions, are based merely on substituting one pragmatic criterion after another, with no reason being given for supposing that any are especially relevant to the project of social illumination. Mirowski's approach seems, from the perspective in question at least, to be quite consistent with Veblen's, and well illustrates the problems that arise from neglecting the project of realist social realizing. It is conceivable, of course, that Mirowski will prove to be correct in his speculations, and even that such an orientation might have some explanatory success. However, if Veblen's speculation (although false as a generalized prediction) concerned a model that has sometimes proven explanatorily successful, despite Veblen's failure to support it on ontological grounds, the same fortunate experience cannot be anticipated for everyone who adopts his neglect-of-ontology orientation.

Finally, let me add a brief word about (ii), i.e., about Veblen's specific emphasis upon institutions and habits, an emphasis resulting directly from his support for an (evolutionary) approach perceived as modern. The danger here is that contemporary institutionalists will interpret institutions and the like as central units of analysis just because of the emphasis they receive in Veblen's writing. This possibility conditions a worry that where the mainstream project supposes in a priori fashion that (atomistic) individuals constitute the main unit of analysis, some institutionalists may, occasionally at least, come to assume that institutions can substitute for individuals in this role. Certainly the possibility of such a reductionism remains a danger wherever ontology is neglected; it is by confronting ontology in an explicit and sustained way that we can be most confident that such errors are systematically avoided. Thus, I think the emphasis of the project of critical realism remains useful for orthodox and heterodox projects alike.

12 Concluding Comments

My basic point in this chapter is simply to emphasize that when ontology is neglected, social theorizing always stands a chance of getting in a bad way. This is most clearly seen to be the case once we examine the performance, practices, and rationales of the mainstream project of modern economics. Indeed, it is clear from such an exercise that the modern discipline is in need of a major overhaul, of a significant reorientation of direction.

Heterodox traditions like old institutionalism clearly fare better, but even here there is a danger that occasionally pragmatic or related concerns (support for approaches merely because they are perceived as modern or Veblenian, etc.) will dominate ontological considerations. Context matters: the nature of the object of study has a bearing on the methods appropriate to its analysis.

In recent years, critical realism has, I think, achieved some headway in the heterodox economics literature, and it is clear that some (old) institutionalists are now recognizing the need for an explicit ontological turn. Of course, it is conceivable, indeed likely, that this turn will not be appreciated by all (old) institutionalists. If so, the result may be a (momentary at least) parting of the ways of different groups within the institutionalist camp (separating on the one hand the realists, more concerned to question the contextual relevance of certain methods, and to criticise where appropriate the contributions of Veblen, etc., and, on the other hand, those institutionalists of a more postmodernist or pragmatist hue, less willing perhaps to allow criticism of Veblen, or perhaps wanting to reduce reliance on formalistic methods to an a priori principle). If this happens, entailing a separation which in any case will doubtless merely reflect an already existing, if usually implicit and largely unrecognized, difference of orientations or emphasis, I believe the outcome will at least be an achievement of greater clarity concerning issues fundamental to the social theory process. Perhaps such (momentary) intellectual division (if combined with charity and respect for the opponent's view) is even essential if the (old) institutionalist project is to forge significant further progress over the short to medium term.

References

Allais M (1992) The economic science of today and global disequilibrium. In: Baldassarri M et al. (eds) Global disequilibrium in the world economy. Macmillan, Basingstoke

Bell D, Kristol I (1981) The crisis in economic theory. Basic Books, New York

Bhaskar R (1978) A realist theory of science. Harvester Press, Hemel Hempstead

Blaug M (1980) The methodology of economics: or how economists explain. Cambridge University Press, Cambridge

Hahn F (1985) In praise of economic theory, the 1984 Jevons Memorial Fund lecture. University College, London

Hayek FA (1942–44) Scientism and the study of society. Economica. Reprinted in The counter-revolution of science. Liberty Press, Indianapolis

Hicks J (1986) Is economics a science?. In: Baranzini M and Scazzieri R (eds) Foundations of economics. Basil Blackwell, Oxford, pp 91–101

Ingrao B, Israel G (1990) The invisible hand. MIT Press, Cambridge

Kirman A (1989) The intrinsic limits of modern economic theory: the Emperor has no clothes. Econ J 99:395

Lawson T (1997a) Economics and reality. Routledge, London

Lawson T (1997b) Situated rationality. J Econ Method 4:101–125

Lawson T (1999) Feminism, realism and universalism. Feminist Econ 5(2):25–59

Leamer EE (1978) Specification searches: ad hoc inferences with non-experimental data. Wiley, New York

Leamer EE (1983) Lets take the con out of econometrics. Am Econ Rev 73(1):34–43
Leontief W (1982) Science 217:104–107 (letter)
McCloskey D (1986) The rhetoric of economics. Wheatsheaf Books, Brighton
Mirowski P (1998) Machine dreams: economic agents as cyborgs. In: Davis JB (ed) New economics and its history: annual supplement to vol 29, history of political economy. Duke University Press, Durham, London
Rubinstein A (1995) John Nash: the master of economic modelling. Scand J Econ 97:9–13
Veblen T (1898) Why is economics not an evolutionary science? Q J Econ vol XII, July. Reprinted in The place of science in modern civilization and other essays. 1919, Viking Press. Republished (with a new introduction by Warren J. Samuels) in 1990 by Transaction Publishers, New Brunswick, New Jersey (page references to the latter)
Veblen T (1908) The evolution of the scientific point of view. Univ Calif Chronicle vol X, No. 4. Reprinted in The place of science in modern civilization and other essays. 1919, Viking Press. Republished (with a new introduction by Warren J. Samuels) in 1990 by Transaction Publishers, New Brunswick, New Jersey (page references to the latter)
Weintraub ER (1989) Methodology doesn't matter, but the history of thought might. Scand J Econ 91:477–493
Wiles P, Routh E (1984) Economics in disarray. Basil Blackwell, Oxford

Author Index

Subject Index

MIX
Papier aus verantwortungsvollen Quellen
Paper from responsible sources
FSC® C105338

If you have any concerns about our products,
you can contact us on
ProductSafety@springernature.com

In case Publisher is established outside the EU,
the EU authorized representative is:
Springer Nature Customer Service Center GmbH
Europaplatz 3, 69115 Heidelberg, Germany

Printed by Libri Plureos GmbH
in Hamburg, Germany